URBAN TRANSPORT

Urban Transport

Edited by
KEVIN HEY
JOHN SHELDRAKE

Ashgate

Aldershot • Brookfield USA • Singapore • Sydney

Published by
Ashgate Publishing Ltd
Gower House
Croft Road
Aldershot
Hants GU11 3HR
England

Ashgate Publishing Company
Old Post Road
Brookfield
Vermont 05036
USA

British Library Cataloguing in Publication Data

Urban transport : a century of progress?
    1.Urban transportation 2.Urban transportation policy
    I.Hey, Kevin II.Sheldrake, John, 1944-
    388.4

Library of Congress Catalog Card Number: 97-72668

ISBN 1 85972 466 3

Printed and bound by Athenaeum Press, Ltd.
Gateshead, Tyne & Wear.

Contents

Figures and tables

Chapter abbreviations

BAA	British Airports Authority.
BRB	British Railways Board.
BTC	British Transport Commission.
CLC	Cheshire Lines Committee.
CRG	Community Reference Group.
DWA	Designated Waiting Area.
GLC	Greater London Council.
GWR	Great Western Railway.
LCC	London County Council.
LGOC	London General Omnibus Company.
LNER	London and North Eastern Railway.
LOR	Liverpool Overhead Railway.
LPTB	London Passenger Transport Board.
LUL	London Underground Limited.
MCC	Metropolitan County Council.
MDHB	Mersey Docks and Harbour Board.
MET	Metropolitan Electric Tramways.
METRAC	Metro Action Committee on Public Violence Against Women and Children.
MPTE	Merseyside Passenger Transport Executive.
NBC	National Bus Company.
OECD	Organization for Economic Co-operation and Development.
PCC	Presidents' Conference Committee.
PSBR	Public Sector Borrowing Requirement.
PTA	Passenger Transport Authority.
PTE	Passenger Transport Executive.
SELNEC	South East Lancashire, North East Cheshire.
SMT	South Metropolitan Tramways.
SPTE	Strathclyde Passenger Transport Executive.

SRA	Special Review Area.
TGWU	Transport and General Workers Union.
TRC	Toronto Railway Company.
TTC	Toronto Transit Commission.
TTPC	Toronto Transportation Commission.
UERL	Underground Electric Railways Company of London.

The contributors

Kevin Hey is Lecturer in the Department of Business Studies at the University of Salford.

Dr. Martin Higginson is Economic Adviser to the Confederation of Passenger Transport U.K.

Dr. James McConville is London School of Foreign Trade Professor and Director of the Centre for International Transport at London Guildhall University.

Stephen Shaw is Senior Lecturer in Transport and Tourism in the Business School at the University of North London.

Dr. John Sheldrake is Reader in Modern History in the Department of Politics and Modern History at London Guildhall University.

Dr. William Tyson is Director of Planning and Promotion at Greater Manchester Passenger Transport Executive.

Acknowledgement

The Editors would like to thank Jan Weekes of the University of Salford for her help and patience in the production of this book.

K.H. and J.S.
Salford and London
January 1997

1 Introduction

The problems of urban transport, rather like the poor, seem always to be with us. Whereas nineteenth century cities suffered from the congestion and pollution caused by horsedrawn traffic, today's equivalents face the threat of grid-lock as increasing numbers of motor vehicles compete for the confined road space of city centres. There has of course been progress. The transport infrastructures of urban areas have become increasingly comprehensive and sophisticated. Similarly, the numbers of people and volumes of freight moving to, through and within towns and cities have steadily increased. Nevertheless there is a sense of malaise if not crisis. Chronic traffic congestion and the associated problems of atmospheric and noise pollution have a dire impact on the quality of urban existence. Travel within cities is most often considered a chore rather than a pleasure while traffic constitutes a constant threat to cyclists and pedestrians alike.

Although the worst problems of urban transport are most often experienced at the heart of major conurbation's, small towns also suffer from the impact of traffic problems, not least the growing numbers of private cars and freight vehicles on their roads. Accommodating private cars has become a major difficulty as available space is now inadequate for the number of vehicles requiring to be parked. The result of this is frustration, loss of amenity and, perhaps, relative commercial decline. Numerous piecemeal measure have attempted to ameliorate the problem, ranging from park and ride schemes to the construction of multi-storey car parks. In some cases the remedy has been worse than the malady.

It is now generally agreed that the present difficulties in urban transport spring fundamentally from the popularity of the motor car. Whatever the personal and social costs of motoring there is no evidence that the era of the automobile is ending. The modest revival in cycling may be an early indication of a return to simpler forms of local transport or merely a passing fad. In any case no one is seriously claiming that the vast majority of people are about to get out of their car and on their bike. Instead, a consensus is forming around the notion of multi-modal, high quality urban transport ranging from cars through heavy rail and light rail system to buses and bicycles. There is now, as the Government recently put it, 'a widespread desire to improve accessibility while, at the same time, minimizing environmental impact'.[1]

The chapters contained in this volume are offered as a contribution to the current debate on urban transport. The bulk of them began their life as papers delivered to a colloquium held at University College, Salford in 1995. The range of issues dealt with is wide, ranging from questions of ownership and network planning to such matters as investment, usage and technological change. In Chapter 2 Sheldrake provides an illuminating examination of the origins and early development of the London Passenger Transport Board (LPTB). The expansion of London, both in terms of geographical area and population, the rapid rise in commuting and motor traffic coupled with unrestricted competition of road passenger transport created a dynamic yet chaotic scene ripe for some form of public transport regulation and co-ordination. The two main providers of public transport in London - the London County Council (LCC) and the Underground Electric Railways Company of London (UERL), which included the London General Omnibus Company (LGOC) - were major competitors but became unlikely partners as they found common cause in campaigning for traffic regulation and co-ordination. As often occurs a few key individuals were crucial to the outcome of developments - in the case of the LPTB it was Lord Ashfield of the UERL and Herbert Morrison as Leader of the Labour group on the LCC and later as Minister of Transport in the Labour Government of 1929-31. In spite of their differing approaches both these figures were sufficiently pragmatic for progress to be made. When initial proposals for a scheme of co-ordination were thwarted it was Morrison who embraced the concept of a public corporation to own and operate London's public transport. The support which Ashfield gave to the proposal proved crucial. Although the idea met considerable political opposition it was in tune with the times surviving the fall of the Labour Government in 1930 and was implemented by the incoming National Coalition Government largely as originally conceived by Morrison. The conclusions which Sheldrake draws from all these events is that the LPTB was a pragmatic response to the political situation and transport problems of London although for Morrison it represented the harbinger of state nationalization. The LPTB quickly gained an enviable reputation as it secured the benefits of economies of scale, standarization and centralization. Nevertheless, the absence of democratic control and lack of competition was disturbing to many and these issues were to remain a continuing source of concern throughout the existence of the LPTB.

The regulation and co-ordination of public transport in the provincial conurbations was very different from that initially adopted in London. As Hey points out in Chapter 3 the formation of the LPTB gave fresh impetus to early efforts to voluntarily establish transport bodies in these areas. None of them came to fruition and, as in London, only direct action from central government secured a structure for public transport co-ordination at conurbation level. The close connection between local government and public road passenger transport eventually led to a situation in which developments for public transport came to be shaped as much by the concerns of local government as those of transport. Two radical individuals in the Labour Governments of 1964-70 played a central role in events leading to the reorganization of local government and public transport. Richard Crossman, Minister of Health and Local Government seized the opportunity in an era of change to establish a Royal Commission to determine the most suitable structure for local government outside London. In the meantime Barbara Castle as Minister of Transport was impatient for change particularly in the main urban areas. She set about establishing passenger transport areas, but confined these to the main conurbations where the transport problems were particularly acute. The areas were to cover an entire urban region but initially they were delineated using a combination of commuting patterns and the boundaries of the local government Special Review Areas (SRAs) established in 1958 under a earlier attempt to examine the structure of local government. Subsequently the transport areas were fully integrated into the reorganized structure of local government in 1974 and their boundaries were made to conform with the conurbation level metropolitan authorities. The end result is that the current definition of transport areas are more a reflection of local government considerations than those of transport, with all the inevitable political struggles which accompany the search for an effective and convenient form of local government. The chapter provides an instructive account of these struggles and argues that if the transport problems in the conurbations are to be effectively tackled there is a need for transport areas which correspond with the catchment area of the problem. Alas areas covering a conurbation and hinterland which make sense in terms of dealing with public transport hardly make for units of local government which would find ready and popular acceptance. The conclusion which Hey draws is that the time has come to re-examine the link between transport areas and the existing metropolitan authorities together with considering the most suitable institutional framework to tackle the worsening transport conditions in our major conurbations.

In contrast to the first two chapters which investigate issues in Britain, Shaw in Chapter 4 examines the connection between municipality and civic pride in the Canadian city of Toronto. His study of Toronto's public transport system over the past century illustrates the acceptance of municipal ownership and the justifications for active involvement of local government in mass transit. The unification, modernization and expansion of the transit network is reviewed in the context of the City Beautiful movement. Whilst the movement attracted considerable support it was also subject to scornful attack but despite this the ideal of a comprehensively planned city with attractive and well run services remained a persistent theme. Shaw chronicles public transport development in Toronto at the turn of the century. Whilst many of the features mirrored British experience subsequent developments mark a distinct divergence with public transport in Toronto increasingly perceived as a means of not only maintaining the vitality of the city region but also in establishing Toronto as a 'world city'. Public financial support was granted to the public transport system only in return for fulfilling a number of economic, social and environmental goals. Whilst proposals for an ambitious urban road programme were seen as the quintessential feature of a modern city they only served to expose the fact that the interests of those living in the older city tended to be different from those in the newer suburban areas. The outcome of the debate which emanated from these tensions made a significant contribution to the emergence of a greater political commitment to support public transport. The 25 year consensus amongst the mainstream political parties supporting a fully regulated and integrated municipal service stands in sharp contrast to the situation in Britain. A notable feature of this period is the way in which 'community politics' has contributed towards recognizing and promoting social equity as a key goal of the public transport system. Three major groups have been at the forefront of these developments - disabled people, women and ethno-racial minorities. After many years of discussion and consultation the voluntary organizations which have campaigned for a more accessible and safer public transport system as a means of improving city life are now accepted as an integral element when public transport developments and improvements are considered. The contribution of the three major groups are examined in turn with particular emphasis upon exploring the ways in which this has contributed to the objective of securing civic pride in a socially and culturally diverse city. In conclusion Shaw notes that, whilst there is some similarity between the current rhetoric of the contribution which public transport can make to this objective with that used by the City Beautiful movement of a century ago, the emphasis is now upon diversity rather than conformity.

In Chapter 5 Higginson examines the policies for investing in city transport over the last 100 years. The key investment periods are identified in terms of the mode of focus and the principle source of investment. One of the most notable features is the continual addition of new transport infrastructure which has occurred alongside the retention of existing facilities. For passenger services these additional facilities have produced an overlay of alternative modes which in urban areas have contributed to the existence of increasingly complex travel patterns over far greater distances than was previously the case. By contrast a different situation has emerged for freight movement which is now primarily inwards from outside urban areas. The contribution of each mode to urban transport is examined in turn. The local networks provided by the railway companies within the largest urban centres were funded by private capital but fare rates and charges were subject to government regulation. When the railways were brought into public ownership the issue of financial assistance became a clouded one. The distinction between capital and revenue support was not always clear and it is only in comparatively recent times that such differences were clarified. After the establishment of the Passenger Transport Executives (PTEs) in the major cities public finance enabled a number of former and, in certain cases, new city centre rail links to be developed. More recently private finance has made a significant contribution to some schemes and with the advent of rail privatization the role of private sector capital is of increasing importance.

By contrast, underground railway development was confined to three cities - London, Liverpool and Glasgow. In all cases initial investment was made by private companies but thereafter developments in each city were different. In London international finance played some part, with public funds used later as part of the initiative to relieve unemployment. Public funding became the source of finance for development of the underground network once the system was brought under the auspices of the LPTB. Whilst the network remains in public ownership, albeit now as a separate company, private sector funding has once again started to play a role in both capital and operational projects. In provincial cities there were different approaches for funding underground railway development. In Liverpool the lines passed into state ownership with the creation of the British Railways Board (BRB) and survived to become part of an improved rail system linking formerly separate suburban railways across the city centre. Public funding for the project came from a number of sources. A different outcome occurred in Glasgow when the underground system passed into municipal ownership and was subsequently transferred to the local PTE under whose control it was comprehensively reconstructed.

The bus industry developed in a different fashion from the railways with a mixture of private and public sector finance. For a long period it maintained overall profitability. The regulatory regime which applied for most of the century permitted the maintenance of an extensive network under which loss making services were supported by profitable ones. With the majority of private companies brought into public ownership the industry became predominantly a public sector activity. As overall profitability declined external financial support became necessary for both capital and revenue purposes. Whilst capital finance was allocated for specified items revenue support was given on a more general basis. Since privatization and deregulation of the industry a decade ago the use of blanket subsidies has been replaced by support restricted to routes or services which cannot be provided on a commercial basis.

The funding of roads is increasingly a source of debate. They have traditionally been funded from central and local government. Once road users are in possession of a road fund licence they have unlimited access to the highway network. The issue of charging for actual road use remains a difficult area. There are a few proposals for privately funded highways whilst the idea of tolling certain roads has also received consideration. Urban road pricing seems as far away as ever despite the case for making road users meet a greater share of the costs they impose but there is little political enthusiasm for such a radical change to charging policy. Higginson concludes that a number of concerns have remained common features during the last 100 years. There is still considerable debate about whether transport is a public good or a commercial business whilst issues of traffic congestion, although different, remain largely unresolved.

The theme of transport development is maintained in Chapter 6 in which McConville discusses the factors influencing the moves towards deregulation and greater market flexibility during the last two decades. The transport sector is fortunate in possessing substantial economic assets mainly in the form of infrastructure which were expensive to develop but have extremely long lives. Historically this raised concerns about the nature of market development with state involvement invariably focused upon control and ownership. By contrast, the mobile units of transport have different characteristics and consequently state attention has been based upon safety and social considerations. The basic economic principles of transport serve as a reminder that time and cost of travel are viewed as the central determinants of transport efficiency. In these terms the motor car is seen by many as a efficient form of transport.

The personal mobility bestowed by the motor car is one of the main ingredients of modern life and it has had profound effects upon work, shopping, leisure and educational activities. The changes in travel patterns which have accompanied the rise in car use has led to a re-emergence of the urban transport problem which is perceived in terms of congestion, parking, loss of social welfare, wasteful use of resources and environmental deterioration. The response of government to this has been through a combination of infrastructure development and providing the organizational structure for the transport sector. To a large extent this approach has been influenced by the perception of researchers, sectional and industrial pressure groups which until recently advocated an engineering response through creating an infrastructure to accommodate the expanding traffic. Whilst this solution was accepted by governments of all shades there were substantial differences in political and economic philosophy between the two main parties. To a large extent these differences centred upon alternative views regarding ownership and competition which was distinguishable through the language used and changes to the organizational framework over the years.

Enthusiasm for this approach began to wane as the financial consequences of the oil crisis and the fracturing of the postwar political consensus impacted upon the transport scene. The engineering solution was increasingly brought into doubt as realization dawned that transport management systems could offer a greater contribution to mobility. This diluted engineering solution emerged in a period of increasing concern for the environment, the scarcity of fossil fuels and major constraints upon public expenditure. This latter element proved the most crucial since it necessitated a reappraisal of the priorities accorded to transport as against other competing claims upon public expenditure, along with a stringent re-examination of the social and economic objectives which were deemed to accrue from such expenditure.

During the last two decades the move towards privatization and deregulation have been undertaken partly on the basis of an ideology which views private ownership and the supremacy of the market as central to efficient public sector finances. Economic growth and consumerism were paraded as the central purpose of life. Yet the trends of deregulation and privatization created a dichotomy in policy making as tension emerged between the competitive solution and increasing public concern with social and economic welfare, particularly in the context of the environment. During this period a number of influences have seen the state, in whatever form, retreat from the role of transport provider to undertake a strengthened one of supervisor and regulator in order to act as guardian of those who consume it.

The discussions in Chapters 5 and 6 provide a firm foundation for the contribution from Tyson who explores metropolitan transport policy in the twentieth century and points to likely developments in the next. Despite the rapid growth of motorized transport it is only in the last quarter century that it has been easy to identify a definable transport policy for urban areas. Prior to this period transport was generally profitable and whilst there was a growing transport problem the response of successive governments was to adopt the engineering solution. In short there was only occasional need for central government to make policy decisions on transport. The laissez-faire attitude of the nineteenth century extended to transport policy in the prewar era. It was only after World War One that significant government intervention occurred with the restructuring of the railways and then regulation to control competition in the road freight and passenger transport industries. This period was characterized by concerns about regulating the level of competition but subject to this protection operators were left to survive in a commercial environment. After World War Two the policy debate shifted to matters of ownership and the structure of the industry with the railways and considerable sections of road passenger and freight transport brought into public ownership.

A radical review of transport policy in urban areas was clearly necessary and this finally occurred as part of a more general review of policy by the Labour Government in the 1960s. This resulted in recognition that the main urban areas were different and special Passenger Transport Authorities (PTAs) were required for the largest conurbations. It was acknowledged that public transport conferred wider benefits such as assisting in reducing traffic congestion and it should not necessarily have to be provided on a commercial basis. The transport authorities were given extensive powers to subsidize public transport as well as co-ordinating bus and rail services. There were other significant developments during this period. Extensive transportation studies were conducted using sophisticated analytical techniques to predict levels of car and public transport use. These resulted in the emergence of a list of infrastructure developments thought necessary to provide for transport demand. The reorganization of local government some years later allowed, for the first time, decisions on matters of transportation and public transport to be considered by one body. Policies for public transport were developed through capital and revenue support with emphasis upon the concept of integrated transport systems as central to attempts to arrest car use. As Tyson rightly observes there was no single policy formula as each area adopted a different combination of investment and subsidy using various modes. Whilst these policies had some success the high costs of subsidy became a matter of increasing concern. Eventually central government intervened to control them and it was a clear indication that policy was starting to turn away from the model of a publicly financed and integrated transport system. The shift of policy towards deregulation and privatization was to occur sooner rather than later and at a time of further reform to the structure of local government in the conurbations. By 1986 responsibility for matters of transportation was, in some areas, fragmented amongst a number of local authorities. Whilst PTAs were retained to deal with issues of public transport the deregulation and privatization of bus operations and the rail network has reduced considerably the scope and influence of these specialist transport bodies.

The transport problems in the main urban centres are continuing to deteriorate and as concern grows about the environmental impact of traffic levels policy makers are giving greater consideration to the transport implications of land-use planning decisions. Tyson provides a timely reminder that the problems of financing public transport have not necessarily been solved by deregulation and privatization. Some doubts are being expressed about the longer term stability of the commercial service network which, if realized, could lead to a growing demand for subsidy in the future.

A number of potential approaches to transport policy are discussed. One response is to continue with a market based philosophy with the introduction of road pricing to ensure that the prices of public and private transport would reflect their relative social costs with the market delivering the optimum allocation of traffic between modes. Notwithstanding the technological and cost difficulties of such a system it would represent a radical policy change requiring a political will which is currently absent. In short the market based approach will not be feasible for a number of years. Discussion on the 'Great Transport Debate' launched by the government and the recent Green Paper makes interesting reading. A number of proposals are outlined which include increasing the powers of local authorities to deal with vehicle pollution, traffic management and encouraging the use of public transport. Private sector funding of investment and operations is to continue and the government will attempt to move towards a more rational pricing policy rather than using regulation to secure environmental improvement. It is too early to establish the extent to which these proposals will become part of government policy but at least there is recognition of a problem which in the metropolitan areas can only be tackled effectively at conurbation level on a multi-modal basis. Possible local policy options include traffic restraint and improving public transport but they both have a number of potential problems and these are discussed in some detail. Tyson takes the view that the missing ingredient is not analysis of the problems but an acceptance that solutions will involve difficult and often unpalatable decisions. There is at least now a greater awareness of the importance of transport and transport policy than at any time in the last quarter of a century. The transport problem in urban areas will not go away and all the signs are that it will continue to worsen. Experience during the twentieth century has been for transport policy to be characterized by a evolutionary rather than revolutionary process and this seems likely to be the case as the next century dawns.

Note

1   Department of Transport, (1996), *Transport the Way Forward*, Cmnd. 3234, HMSO, London.

2 The monopolization of London's transport

John Sheldrake

Introduction

This chapter provides an account of the origins and early development of the London Passenger Transport Board (LPTB). The section which follows deals with the demographic, industrial, geographic and organizational background. This is followed by a section which covers the various institutional, operational, commercial and political considerations involved. This is followed in turn by a section dealing with the establishment of the LPTB, including legislative and financial arrangements. The final section contains some brief conclusions.

Background

In 1914 London occupied a circle of radius 6 to 8 miles from Charing Cross. By 1939 this had expanded to over 12 miles (Thomas, 1970, pp.23-24). Between the wars the population of Greater London increased by over 1.5 million - rising from approximately 7.25 million to 8.75 million (Barker and Robbins, 1976, p.2). However, as the metropolis spread outwards the population of the inner core declined. Between 1901 and 1939 the number of people living in the London County Council's (LCC) administrative area fell by half a million from 4.5 million to 4 million and this decline continued, reducing the former county's population to under 3 million by 1969 (Barker and

Robbins, 1976, p.2). In the inter-war period suburban London spilled out over the surrounding counties, turning former villages into busy commuter centres and planting industries on what would now be called 'green-field' sites. In part this outward movement contained an element of planning. The LCC, for instance, pursued an active, if sometimes reluctant, policy of 'spreading the people' which was aimed at relocating the inhabitants of inner urban slums to healthier areas where land for building was available and cheap. Whether those living in the 'over-spill' areas welcomed the arrival of the newcomers (or whether the newcomers wanted to relocate) is, of course, another story. More often, in any case, the development of suburban London was unplanned and occurred as a result of speculative enterprize by builders. As Jackson has observed 'by the end of 1938, private firms had finished 618,571 houses in Greater London since the end of World War I, handsomely beating the local authorities' 153,188', (Jackson, 1991, p.64).

Although London's economy to some extent shared in the general economic depression of the inter-war years it was insulated from its worst effects by the diversity of its industrial base.[1]   Again in the words of Jackson 'London escaped the worst. From such setback as it did suffer, recovery was rapid, assisted by the continuing prosperity of the consumer durable industries, the expanding distributive trades and uninterrupted growth in service and administrative jobs' (Jackson, 1991, p.63). The availability of jobs in Greater London, together with housing at a price, stimulated migration into the region from elsewhere in Britain.   During the 1930s, in the words of Barker and Robbins:

Seven new London inhabitants had migrated in for every three produced by indigenous natural increase. London was prosperous and had little difficulty in absorbing these migrants into work in the north, north west, and west along the arterial roads, and into the service industries and trades. Between 1923 and 1939 the total amount of employment in Greater London ... rose from just over 2 million to just under 3 million (Barker and Robbins, 1976, p.3).

Outside of its declining inner core, London was a boom town during the 1930s and this was reflected in increased use of public transport. Commuting to and from work, together with the expansion of leisure activities, generated demand for transport. The rise in real incomes sustained demand and this was met by many improvements in services. Furthermore, technical change enhanced the quality of transport and extensive mobility within the metropolis was taken for granted by its inhabitants. As Jackson has commented 'passenger journeys originating in the Greater London area (trams, buses and Underground and suburban railways) rose from 1.8 million in 1911 to 4 million in 1935, a growth of 120 per cent in a period when population increased by 19 per cent' (Jackson, 1991, p.64). However, having said all of this, the pace of increasing usage slowed from the mid 1930s, as prosperity stimulated the extensive ownership of relatively cheap, mass produced cars. Private housing on the periphery of the metropolis was now often provided with garage facilities and the new arterial roads began to carry ever greater levels of traffic.

As early as 1935 Frank Pick, in his role as vice-chairman of the LPTB, was issuing a warning about the growth of motoring. In a lecture at the Royal Society of Arts he made the following comments:

> There are now about 415,000 motor cars licensed in London and the five Home Counties, or one to every twenty-seven people resident in these areas. All but a small proportion of these motor cars must operate in and about London. They are an extravagant form of transport if user and street occupation are related together. They represent a menace to communal transport. They often cause an abuse of street accommodation. They are the main ingredient in congestion, and it is questionable whether central London can economically support them ... It is a problem demanding renewed study (Pick, 1936, p.213).

Self-evidently it was a problem demanding study and it remains so. Ironically, at the moment in the 1930s when the demise of horses and trams from the streets of London was apparently poised to reduce the capital's notorious traffic congestion, a new, and as we now know, greater menace was appearing in the form of the car. Whereas there were 292,000 private cars licensed in the London Transport area in 1933, by 1938 the figure had already increased to 475,000 and (following a brief decline during World War Two) went on to reach 2,285,000 in 1969 (Barker and Robbins, 1976, p.12).

Implicit in what Frank Pick had to say concerning the increasing ownership of cars and its impact upon public transport and traffic congestion, was a demand for greater planning. In this Pick was very much in tune with his times, for the 1930s were years during which the advantages of planning in all its forms were advocated like never before. Dissatisfaction with the status quo manifested itself in a vast literature inspired by such disparate elements as Soviet Communism, Fascist Italy and Nazi Germany. Equally, calls for greater planning (and an enhanced role for the state within the planning process) came from parliamentary figures such as Lloyd George, Hugh Dalton and Harold Macmillan. Support for planning brought with it the advocacy of regulation, co-ordination and control. Management of complex, technological societies required expertise. Technocratic values tended therefore to displace democratic ones. As will be seen subsequently the LPTB was a case in point.

In any case London had long been an example of the trade off between efficiency of operation and democratic control. To this extent the capital has always suffered from what would now be termed the 'democratic deficit'. The existence of the historic City of London at the heart of the metropolis was (and perhaps remains) an anomaly. Throughout the nineteenth century it remained opposed both to reform and expansion. As a result London did not share in the municipalization movement which characterized the development of, for example, Liverpool, Manchester and Birmingham. Instead it had to make do with a combination of adhoc bodies and vestries topped off, after 1855, with the indirectly elected Metropolitan Board of Works, primarily created to provide the metropolis with an adequate system of sewers. Only in 1889 with the establishment of the LCC did London obtain a democratically elected, top tier local authority on the modern pattern. Even so the LCC was lacking in powers compared to its provincial counterparts. It never gained control of London's water supply which was placed in the hands of the Metropolitan Water Board. Equally it never controlled London's extensive port facilities which were overseen by the Port of London Authority. Finally, it did not control either the capital's gas industry (which remained in private hands until nationalization) or electricity supply (which was shared between private companies and the metropolitan boroughs - i.e. the lower tier authorities).

Only in the sphere of public transport did the LCC gain control of a major public utility - namely the tramway system. Initially lacking the powers to operate the trams itself it had to wage a long campaign before it eventually acquired the dozen or so private tramways operating in its area. During the 20 years this process took the horse tram was increasingly giving way to electric traction, although this change was slow to come in London. Electrification of the LCC's system was not substantially achieved until 1915 and only completed in 1921. By this latter date the notion that owning and operating London's trams would in some way provide the LCC with the control of London's public transport was already passé. The Council's trams faced increasing competition from buses and Underground railways, neither of which did it operate or control. By the early 1920s the expansion of the metropolis far beyond the LCC's administrative boundary (together with the operations of buses, Underground and, indeed, suburban railways) had already rendered it impossible for the Council to control, operate or plan London's public transport system without a massive reorganization. Instead its activities were limited to making common cause with its major competitor in the campaign for traffic regulation in the early 1920s which resulted in the passage of the London Traffic Act, 1924.

Origins

By the mid 1920s there were two major providers of public transport in London - namely the LCC and the Underground Group. A third might be added if the extensive (and increasingly electrified) commuter network of the Southern Railway were included. The LCC tramway system has already been mentioned. Although, like all tramway operators, the Council was burdened with the expense of road repairs under the Tramways Act, 1870, and was also providing cheap workmen's tickets, its tramway operation remained a viable project. The London Traffic Act, 1924, gave a modicum of protection from the worst depredations of private bus operators and the Council also attempted to meet the challenge of bus competition through renovation of its fleet. As Wiseman has commented:

> The LCC ... had to face up to intensive omnibus competition and it met this with the "Pullman" programme of renovation. Between 1926 and 1929 most E1 cars (the Council's standard vehicle which appeared in 1907 and eventually totalled 1,050 units) were fitted with comfortable upholstered transverse seating etc. They were painted in a new red livery in place of the traditional chocolate and by 1933 the new livery was applied to all LCC cars then extant (Wiseman, 1986, p.83).

In 1930 the Council introduced new trams to replace the oldest E1s and, by 1931, 260 of the new HR2s and E3s had been added to the fleet. When, in 1933, the LCC's tramways were transferred to the LPTB the fleet consisted of 1,713 cars. Unlike many provincial tram operators the LCC had not begun the process of replacement by trolleybuses. Although the Council had experimented with vehicles and had plans to provide trolleybus routes in south east London 'all schemes failed, largely because the local councils were granted a power of veto to bolster up their customary objections, and Londoners had to suffer gladly, as usual, in the interests of prejudice' (Newman, 1953, p.130). Further, the Council lacked the powers to operate buses - all attempts to obtain such powers being resisted by the capital's bus interests. In this situation perseverance with the tramways was more or less inevitable. Certainly, if Wiseman's view is anything to go by, the LCC tramways were not in a state of decay during the early 1930s. As he put it, 'there is no doubt that when the LCC system was transferred to the LPTB ... it was at the height of its efficiency' (Wiseman, 1986, p.85).

The Underground Group operated tramways from outside the LCC's administrative area and these were combined in the London and Suburban Traction Company, consisting of Metropolitan Electric Tramways (MET), London United Electric Tramways (LUT) and South Metropolitan Tramways

(SMT). None of these operators possessed large fleets. At the time of transfer to the LPTB the MET contributed 316 cars, the LUT 150 and the SMT 52 (Wiseman, 1986, p.75). Like the LCC the privately owned tramways were active in the late 1920s trying to upgrade their vehicles in order to meet the competition from buses. During 1929 and 1930 a prototype 'Feltham' tram was built for the MET by the Union Construction and Finance Company (a subsidiary of the Underground) at Feltham in Middlesex. A 100 'Felthams' were built in 1931, bringing 'a new dimension of speed and comfort to the London transport scene' (Wiseman, 1974, p.80). Whereas the LCC had been prevented from undertaking the deployment of trolleybuses, private companies succeeded in running the new vehicles. As Barker and Robbins have commented 'London United ... seriously got to work with conversion of its south westerly routes ... to trolleybus operation ... The first trolleybus service in London, between Twickenham Junction and Teddington began on 16 May 1931 - inaugurating a comparatively short era' (Barker and Robbins, 1976, p.241). The vehicle used by LUT was built by United Construction at Feltham and resembled the 'Feltham' tram.

Having said all of this, however, suburban trams and trolleybuses formed a very modest part of the Underground Group's road passenger transport operations. The bulk of its activity was based on the buses provided by the London General Omnibus Company (LGOC) and its associate companies which owned some 5,000 vehicles. From the early 1920s the LGOC had fought a long battle against the independent (so called 'pirate') operators which they had more or less won through a combination of company acquisitions and regulation under the London Traffic Act, 1924. When, in the late 1920s, the LGOC came under fresh pressure (this time from coach operators running vehicles into London from beyond the Metropolitan Police District) they responded by establishing Green Line in July 1930. Although the growth of Green Line was constrained by the regulatory provisions introduced under the Road Traffic Act, 1930, it nevertheless boasted a fleet of over 400 vehicles by 1932 (Warren, 1980, p.18; Barker, 1990, p.90). Finally, of course, the Underground Group had at its core the extensive network of tube railways which, notwithstanding their history of financial under-performance, succeeded during the 1920s in extending their tentacles ever deeper into the London suburbs. The Underground Group was sustained by an elaborate system of cross-subsidy between services and modes and supported, as has been seen in the case of the 'Feltham' trams and the LUT's trolleybuses, by its own equipment manufacturers. When challenged the Group (or Combine as it was increasingly called) was prepared to compete and to acquire. However, its preferred trading position was one of strictly managed competition or, better still, qualified monopoly.

Two remarkable men presided over the operation of the Underground Group - namely Lord Ashfield (formerly Sir Albert Stanley) and Frank Pick. Ashfield was born in Derby in 1874 as Albert Knattries. When he was 11 his family emigrated to Detroit where his father changed the family name to Stanley. The young Albert Stanley began his transport career at the age of 14, working in the stables of the Detroit Street Railway Company. By the time he was 28 he had risen through the ranks to become general superintendent. In 1904 Stanley left Detroit and became superintendent of the railway department of the Public Service Corporation of New Jersey. In 1907 he became general manager of the Corporation. At this time the finances of London's Underground Electric Railways (UERL) were in a sorry state and American shareholders 'head hunted' Stanley to sort out the problems. He accepted and was startlingly successful. As Barker has commented 'he brought the business together and saved it from disaster, becoming managing director in 1910 and then ... acquiring the London General Omnibus Company, the Central London Railway and other transport undertakings' (Barker, 1990, p.81). In 1914 Stanley was knighted for his services to London's passenger transport and in 1916 joined Lloyd George's first government as President of the Board of Trade. In 1919 he returned to what was now called the Underground Group as Chairman - a post he retained until the LPTB was established in 1933. In 1920 he received a peerage, taking the title Baron Ashfield of Southwell. In 1923, in the context of the campaign which led to the London Traffic Act, 1924, Ashfield spelt out his views on the management of London's transport in the following terms:

What is needed is some responsible and judicial authority, able to say what is required stage by stage, for the development of London's traffic facilities ... Every metropolitan city has been compelled to this solution. It is the case in New York, in Paris, and in Berlin. Since the war circumstances have hastened all these cities to consolidate their traffic undertaking, and to ensure their future growth and development come by design and not by accident. Competition is a dangerous weapon. It may seem to offer immediate gains, but they are at the expense of future losses. This is the universal experience (quoted in Barker and Robbins, 1976, pp. 208-9).

Like Ashfield, Frank Pick was a long term servant of the Underground Group. He was born in Spalding in 1878 and educated at St Peter's School, York. After leaving school he was articled as a solicitor, qualifying in 1902 and obtaining a London LLB in 1903. In 1902 he started work for the North Eastern Railway Company, joining the staff of its general manager, Sir George Gibb. In 1906 Gibb moved to London to take over the management of UERL and took Pick with him. When Gibb retired from direct management responsibility in 1907 Pick was transferred to the staff of his successor, Albert

Stanley. In 1909 Pick became traffic officer of the UERL and, following the LGOC takeover, was appointed commercial manager in 1912. In the latter job he had particular responsibility for building up the network of bus routes and also for advertizing. Pick's efforts in the sphere of advertizing technique are well known and he succeeded in turning the Underground Group into a leading patron of poster art and, as will be seen in the next section, industrial design in general. He succeeded in providing the Underground Group with what we would now call a strong brand image manifested in such artefacts as the company logo and the ubiquitous Underground network map. As Green has commented:

> Pick reserved special illuminated boards at the Underground station entrances for the company's own pictorial posters and maps ... The poster grids on the station platforms were separated from the station nameboards which, after 1908, appeared in an early version of the bar and circle symbol with the station name in white on a blue bar across a red disc ... The typography of the posters also required special attention. Pick was not happy with the traditional typefaces used by the printers, and decided that the Underground needed its own display typeface to distinguish the company's information clearly from the other commercial advertizing on its property. The eminent calligrapher, Edward Johnston, was commissioned to design the new lettering ... and a modified version known as "New Johnston" is still the standard typeface applied to the Underground's posters today (Green, 1990, pp. 9-10).

In 1917 Pick was appointed by Sir Albert Stanley (in his role as President of the Board of Trade) to take charge of the household fuel and lighting branch of the coal mines control department. Like Stanley he returned to the Underground Group after the war, becoming joint assistant managing director in 1921 and taking full administrative control under Ashfield in 1924. In 1928 he became managing director of the Underground Group and president of the Design and Industries Association. The abilities and interests of Ashfield and Pick were complementary. F.A.A. Menzler (Chief Development and Research Officer for the London Transport Executive) who worked for many years with both men, made the following observations about their respective spheres of activity within the Underground Group and subsequently the LPTB:

> Lord Ashfield as Chairman was, in a way, above the battle. His position was, so to speak, Ministerial ... Roughly speaking, he concentrated on finance, policy, and external affairs ... Pick, though naturally fully consulted on all important matters, conducted the day-to-day management of the vast complex of services ... The chairman left the details of organization mainly

to Pick. Contrary to widespread belief, Lord Ashfield, at any rate during the years I knew him, was not remarkable in matters of organization: Pick seemed to do it all ... With such a colleague, Lord Ashfield was relieved very largely of the burdens of the day-to-day management of the undertaking. He could sit back with a clear desk in a large, handsome room with not a paper in sight and so to speak, "think imperially" (Menzler, 1951, p.102).

As well as providing a framework for the regulation of road passenger transport in the metropolis, the London Traffic Act, 1924, established the London and Home Counties Traffic Advisory Committee. The role of the Committee was to provide advice to the Minister of Transport on metropolitan transport issues and, as Barker and Robbins have observed, 'the reports it produced in the later 1920s were extremely important in influencing ownership and operation of London's transport' (Barker and Robbins, 1976, p.210). In 1927 it produced the so called Blue Report, *A Scheme for the Co-ordination of Passenger Transport Facilities in the London Traffic Area*, which proposed 'the co-ordination, joint management, and pooling of receipts of bus, tram and underground railway passenger transport services' (Klapper, 1978, p.109). It also advocated the retention of existing patterns of ownership. As Donoughue and Jones have commented:

The proposals of the Advisory Committee pleased Ashfield, who in the middle of 1926 met the leaders of the Municipal Reform majority of the LCC. Both sides were in agreement ... (and) the Combine and the LCC sought ministerial blessing for their agreement. Having a large legislative programme the government was unwilling to promote extra legislation unless it was non-controversial ... The Combine and the LCC decided to proceed by private legislation, which would require two bills, one relating to the powers of the LCC and the other to the Combine. By the autumn of 1928 the bills were ready. They enabled the Combine and the LCC to adopt a scheme of co-ordination, establishing a joint management structure and common pool of receipts but retaining separate ownerships (Donoughue and Jones, 1973, p.121).

By May 1929 both bills (i.e. the London County Council - Co-ordination of Passenger Traffic Bill and the London Electric Railways - Co-ordination of Traffic Bill) had completed their second reading in the Commons and would no doubt have been passed by the Lords. However, their final passage through the Commons was prevented by the dissolution of parliament, the return of a Labour Government and the appointment as Minister of Transport of Herbert Morrison.

Morrison, in his role as leader of the Labour group on the LCC, had along been a critic of the Combine and what he saw as Ashfield's attempts to gain control of the Council's tramway system on the cheap. He had opposed the London Traffic Act, 1924, even though it was enacted by a Labour Government, and had clashed with Ernest Bevin the leader of the Transport and General Workers Union (TGWU) on the issue of representation on the Advisory Committee. Whereas Bevin was ever eager to extend the influence of the TGWU by adopting a bi-partisan corporatist position, Morrison (at least in 1924) remained committed to the extension of local government influence. Although the two men were subsequently associated for many years in Labour Party affairs this apparently minor disagreement was never forgotten. In the words of Donoughue and Jones, 'by standing up to Bevin, Morrison earned his everlasting hatred ... For the rest of his life Bevin was to pursue Morrison with venom' (Donoughue and Jones, 1973, p.120). Having killed off the two Co-ordination Bills, Morrison set about providing an alternative structure. Ironically considering his local government credentials, Morrison turned to the idea of a public corporation on the model provided by the Central Electricity Board and the British Broadcasting Corporation. After giving preliminary consideration to Morrison's proposals, the Cabinet established a sub-committee under Morrison's chairmanship to further examine the viability of his proposals. In November 1929 the sub-committee reported to the Cabinet in the following terms:

It first advised that the area of the authority should extend over a radius of 25 miles from Charing Cross, the so-called London Traffic Area of 1924. It listed 5 undertakings to be brought in: the railways, omnibuses and tramways of Ashfield's Combine; the Metropolitan Railway ...; the tramways of the LCC and other local authorities; the suburban parts of the mainline railway companies; and the omnibuses of other private companies (Donoughue and Jones, 1973, p.142).

With the exception of the suburban railways which were to remain beyond its direct control, this describes the geographic range and operational scope of the LPTB when it was established in 1933.

The plan to nationalize Lord Ashfield's Combine was, on the face of it, an audacious one. However, Ashfield was by no means opposed and, in lengthy negotiations with Morrison and his officials, exercised his greatest efforts not in resisting a public corporation but by obtaining the best price for his shareholders' assets. For his part Morrison found Ashfield to be 'a man he could do business with'. Again in the words of Donoughue and Jones, 'Morrison came to admire Ashfield and had him in mind to be the chairman of the new board. To nationalize Lord Ashfield was his objective. Morrison had a high regard for his efficiency, his spirit of public service and his friendly relations with the unions' (Donoughue and Jones, 1973, p.145). Morrison was equally impressed with Frank Pick and 'advised that he be appointed to the new board as deputy to Ashfield. Pick was nationalized too' (Donoughue and Jones, 1973, p.145). While Ashfield was engaged in convincing his shareholders that the creation of the LPTB was in their interests, Morrison set about persuading the sceptics among his Labour Party colleagues to accept the idea of a public corporation to run London's transport. The ramifications involved in this were significant in that the Party's previous calls for nationalization had been couched in somewhat vague and idealistic terms. Morrison's proposals, in contrast, were couched in concrete terms and came to set a precedent for subsequent nationalization's carried out in the years after 1945. As Morrison observed in 1933:

My own Party had never worked out its socialization proposals in Government Bills ... in the days of the second Labour Government its ideas were by no means clear. There was the view represented by the earlier assumptions of ordinary State Department nationalization; there was that of orthodox municipalization, or management by municipal joint committee; there were the vague ideas about workers' control, Guild Socialism, and some even bordering on Syndicalism. We had to fit platform speeches, vague Party declarations, and the actual facts, into a detailed Parliamentary Bill to be promoted on the responsibility of the Government (Morrison, 1933, p.114).

In the event Morrison's bill for the establishment of the LPTB was not published until March 1931. As Barker and Robbins have commented:

The bill was, in parliamentary language, a "hybrid", introduced as a public bill, being a government measure which affected private interests of particular persons or bodies, it had to go through private bill procedure, by which it was subjected to detailed scrutiny, with legal representation for all parties affected and examination and cross-examination of witnesses (including the minister), before a joint select committee of both Houses. The committee first sat on 28 April 1931, facing 80 petitions deposited against the bill, and its work began in earnest on 12 May. The proceedings occupied 35 days in all; the record of the deliberations runs to 1,300 pages of small print - at least a million words (Barker and Robbins, 1976, p.272).

On 20 July the committee's chairman, Lord Lytton, announced that the bill should proceed. By this time, however, political events of great significance served to stop the bill in its tracks. The aftermath of the Wall Street Crash had placed immense pressure on the world economy. Economic slump placed even greater pressure on Britain's already hard hit staple industries, generating increasing unemployment. The determination of Chancellor of the Exchequer, Philip Snowdon, to defend the gold standard and sustain a balanced budget, made a difficult situation even worse. In August 1931, Snowdon, together with Prime Minister Ramsay MacDonald and other leading members of the Labour Government (such as Arthur Henderson and J.H. Thomas) decided on a package of public expenditure cuts as a means of balancing the budget. Against a background of severe financial crisis, with gold flowing out of the country as foreign investors reacted to the fear of a British banking collapse, Snowdon and MacDonald succeeded in obtaining only a bare majority in Cabinet for their proposed cuts. MacDonald's response was to resign on the morning of 24 August 1931 and, apparently under pressure from King George V, become leader of a National Coalition Government. In October 1931 MacDonald led his new government into a general election in order to obtain a mandate for their policies. As Pearce and Stewart have commented 'his message seemed more attractive than that presented by Labour. The National Coalition Government was given its "mandate" with a staggering return of 556, only 13 of them Labour, and took office for the next 9 years' (Pearce and Stewart, 1992, p.268).

Herbert Morrison resigned from office and was replaced as Minister of Transport by the Liberal, P.J. Pybus. Given Conservative hostility to Morrison's Bill, and the strength of Conservative representation within MacDonald's national coalition, the demise of Morrison's nationalization plans seemed the likeliest outcome. In the event 'Pybus ... announced on 6 October not merely that the bill would be carried over to the new Parliament but also, very remarkably, that it would go forward from the stage at which it had been left in July' (Barker and Robbins, 1976, p.273). Some changes were, however, made to Morrison's bill in order to placate Conservative opposition. Regarding the selection of the board, for example, Pybus replaced ministerial appointment with five appointing trustees - namely the Chairman of the LCC, the President of the Law Society, the Chairman of the Committee of London Clearing Banks, the President of the Institute of Chartered Accountants and the Chairman of the London and Home Counties Traffic Advisory Committee. Although Morrison opposed this move, he was nevertheless sufficiently pragmatic to applaud the eventual outcome. As Donoughue and Jones have commented:

Soon after the London Passenger Transport Board came into operation his criticism gave way to praise. The appointing trustees selected members of the board "as I intended", with Lord Ashfield as Chairman and Frank Pick as his deputy; the other members were John Cliff, Assistant Secretary of the TGWU, one each from the LCC and Surrey County Council and one from the London and Home Counties Traffic Advisory Committee ... And as the board embarked on major developments, such as tube extensions, the substitution of trolleybuses for trams, improvement of junctions between tubes and suburban lines and electrification of suburban lines Morrison gleefully pointed out that he had been responsible for setting up the organization which made such developments possible (Donoughue and Jones, 1973, pp. 187-88).

Without doubt Morrison did play a significant part in the establishment of the LPTB but crucial support for the initiative came from Lord Ashfield. It was his support that made a proposal that smacked of socialism more generally acceptable. As Barker and Robbins have observed:

> The Underground Group held the key to the whole operation. Determined opposition from 55 Broadway (the corporate headquarters of the Combine) would have made Morrison's objective virtually impossible to achieve; but Lord Ashfield and Frank Pick had consistently been saying that a common management was an essential element in any solution. Starting from opposed convictions, Morrison the Cockney Socialist and Ashfield the businessman had decided that a single agency was the only answer, and both were sufficiently flexible to adapt enough to each other's approach to make community of view possible. Perhaps Morrison moved further than Ashfield (many people in the Labour Party thought so at the time and subsequently); but without Ashfield the thing could never have been done (Barker and Robbins, 1976, p.273).

Two examples of Ashfield's intervention are worthy of notice. In May 1931 a meeting of stockholders and shareholders was convened to consider the terms which Ashfield had negotiated with Morrison for the acquisition of the transport interests of the Underground Group. Ashfield gave a formal presentation laying out the basis of the offer and then threw the meeting open to general discussion. The sense of the meeting that emerged was firmly against disposing of the Group's assets to a public corporation either on the terms offered, or indeed any terms whatsoever. However, shortly before a vote was taken, Ashfield made a personal appeal for support in the following terms:

> If you have relied on my recommendations in the past, I must ask you to rely upon my recommendations to you this morning ... If you are going to have a public board to deal with (co-ordination) - and I suggest to you that that is the only way it can be dealt with - in my opinion this is the very best form of public Board that can be suggested ... I am saying to you that I have pledged my word to the Minister that I will support it. You may fail to

support me; and in that event you will have to find someone else who will carry on your undertakings for you. I will not pledge my word and then go back on it. The decision is now with you (quoted in Menzler, 1951, p.106).

In the event the shareholders did not call Ashfield's bluff and the necessary majority was gained for the Morrison/Ashfield scheme. Similarly, when the Bill to establish the LPTB finally came to the House of Lords in 1933, Ashfield made his single speech in that chamber in support of the proposals. Again in the words of Menzler, Ashfield's 'speech made a tremendous impression on their Lordships. Until he spoke, there was some doubt as to whether they would pass the Bill, but his speech rendered its passage comparatively easy' (Menzler, 1951, p.106).

Establishment

The London Passenger Transport Act, 1933, gave the LPTB a monopoly over road passenger transport within an area of 1,550 square miles and a total area of operation of some 2,000 square miles. Such was its scale of operation that the LPTB provided a coherence to the notion of Greater London that had not previously existed and became influential in future planning decisions concerning the nature of the capital. As Frank Pick observed in 1934 it covered 'the whole of the territory within 18 miles of Charing Cross and the major part of the territory up to almost 35 miles from Charing Cross. London has at last received a fairly liberal interpretation, which will afford a basis from which to realize the scale of its many problems of planning, housing, public services and so forth'.[2] Of course, the metropolis did not obtain an extension of the LCC until the creation of the Greater London Council (GLC) under the London Government Act, 1963, and even then this body's existence was short lived being abolished in the mid 1980s. In 1933 there was no prospect of a strategic authority being created beyond the LCC's boundaries and this provided sufficient justification for Herbert Morrison to enable him to accept a passenger transport authority lacking in local democratic control.

As has been seen the core of the LPTB's 'hardware' was made up of the bus, tram and tube railway operations of the Combine and the tramway network of the LCC. To this were added the modest tramway networks of the various municipalities within the board's designated area, including Barking, Bexley, the City of London, Croydon, Dartford, East Ham, West Ham, Erith, Ilford, Leyton and Walthamstow. Although the LCC acquiesced somewhat reluctantly in the loss of their trams 'the other tramway-owning municipalities did not prove so difficult - some of them indeed were glad to get rid of these burdens on the rates almost at any price' (Barker and Robbins, 1976, p.277).

Even the LCC, which regretted the loss of democratic control over London's transport, was nevertheless reconciled to the financial outcome. In 1934 Councillor Samuel Gluckstein, chairman of the Council's Finance Committee, was pleased to report that 'the Council now held preferential fixed interest bearing stock with an assured income of more than £100,000 above the sum required to meet the interest on the outstanding tramway debt'.[3]

The greatest resistance came from the Metropolitan Railway which claimed to be quite distinct from the proposed LPTB's tube railway operations and thus sought to be treated as a mainline railway and left alone. A strong argument against this position, however, was the company's operation of part of the Inner Circle tube line. As Jackson has commented:

> The company's attitude, made clear both to ministers and shareholders towards the end of 1930, and subsequently, was that it considered itself to have the same status as the four grouped mainline companies and like them, should be allowed to continue independently ... But the argument did not wash with the Ministry of Transport and its advisors: in London itself, particularly in its equal partnership with the Metropolitan District Railway in the operation and ownership of the Inner Circle, the Metropolitan appeared not greatly different from that company; ... Finally, Lord Ashfield, a powerful influence in favour of co-ordination, was anxious to see the Metropolitan part of a single undertaking (Jackson, 1986, p.289).

Somewhat to the LPTB's chagrin it did not gain its desired control of the capital's suburban railways. Instead a pooling arrangement was made whereby 'the four mainline companies agreed to a pool for all traffic in the London area, the proportions of which provide the best pointer towards the relative strengths of the new LPTB and the London suburban services of the mainlines: LPTB 62 per cent and mainlines 38 per cent, of which the lion's share, 25 per cent, went to the Southern Railway' (Barker, 1990, p.81). Finally, a standing joint committee of the Board and the four mainline companies was established, to consider their mutual interests and to endeavour to settle conflicts between those interests as they arose.

By the time the LPTB came into existence the political and economic climate was substantially different from that which had prevailed in 1931. The National Coalition Government had taken the pound off the gold standard and adopted a policy of qualified protectionism. Interest rates were low and a building boom had begun. Government subsidies to industry and agriculture were now accepted as a legitimate means of reducing unemployment. As has been seen London did not experience the high unemployment levels suffered, for example, in the distressed areas of the North and North East. Nevertheless, the capital's transport infrastructure certainly benefited from the new direction of public policy. Aware 'of the contribution that new railway works made towards the relief of unemployment, government credit was made available to the railway companies and the LPTB' (Jackson, 1991, p.173). Under the London Transport (Finance) Act, 1935, the government guaranteed a loan of £40 million, at a preferential rate of 2.5%, for the improvement of the capital's transport network. Much of the legacy of this expenditure is still visible, particularly in terms of improvements to the Underground network where Frank Pick was able to maintain his influence on design. Further, the LPTB set about the task of replacing what they considered to be the obsolete tramway network with trolleybuses.

Conclusions

The establishment of the LPTB was a pragmatic response to the transport needs of the metropolis. For Herbert Morrison, of course, the LPTB was the precursor of state nationalization and he devoted a lengthy book to the subject (Morrison, 1933). On reflection, however, the LPTB manifested as many continuities (not least the sustained influence of Ashfield and Pick) as it did changes. Certainly the lack of democratic control was worrying to many and the lack of competition of concern to many others. Indeed, as early as 1934, in a lecture at the London School of Economics, Frank Pick found it necessary to claim that although competition from various providers had been ended through the imposition of monopoly, a spirit of competition nevertheless existed within the organization which would serve to keep the enterprize efficient. In 1934 the LPTB was employing some 72,000 staff (a massive organization by any standards) and was constrained to operate as a bureaucracy. Of course, as anyone who has worked in a bureaucracy knows, competition for resources within an organization is a very different thing from the challenge of external competition from alternative providers. With hindsight the LPTB can be placed in the context of the rationalization movement which gathered influence in British political and industrial circles during the late 1920s and early 1930s. The doyen of the movement was the management thinker, Lyndall Urwick.[4] On the eve of the postwar Labour Government's nationalization programme Urwick wrote the following:

> Experiments in public ownership uninformed by the consciousness of management ... will ... make the worst of both worlds. They will destroy the initiative, the freedom to experiment, which are the main virtues of private industry without substituting any alternative dynamic. Unless those responsible are not only convinced that management matters, but are much more clearly informed than at present as to management methods and

principles, they will create not an active and enthusiastic industrial democracy, but a wooden bureaucracy (Urwick and Brech, 1953, p.234).

Morrison of course was confident that public ownership would be both equitable and efficient and certainly in the early days of the LPTB the benefits reaped by economies of scale, standardization and centralization produced an organization which Philip Bagwell has justifiably described as 'the envy of many other countries' (Bagwell, 1986, p.258).

Notes

1   For a recent discussion see M. Daunton, (1996), 'Industry in London:
      Revisions and Reflections', *The London Journal*, Vol. 21 No. 1.
2   *The Transport World,* Vol. 75, 3 March 1931, p.119.
3   *The Transport World,* op. cit., p.123.
4   For a recent discussion see Sheldrake, J. (1996), *Management Theory:
      Taylorism to Japanization,* International Thomson, London, ch. 10.

References

Bagwell, P. (1986), *The Transport Revolution 1770-1985,* Routledge, London.
Barker, T. and Robbins, M. (1976), *A History of London Transport: Volume 2
   - The Twentieth Century to 1970,* Allen & Unwin, London.
Barker, T. (1990), *Moving Millions: A Pictorial History of London Transport,*
   London Transport Museum, London.
Donoughue, B. and Jones, G. (1973), *Herbert Morrison Portrait of a
   Politician,* Weidenfeld & Nicolson, London.
Green, O. (1990), *Underground Art: London Transport Posters 1908 to the
   Present,* Studio Vista, London.
Jackson, A. (1986), *London's Metropolitan Railway,* David & Charles,
   Newton Abbot.
Jackson, A. (1991), *Semi-Detached London: Suburban Development, Life and
   Transport 1900-39,* Wild Swan Publications, Didcot.
Klapper, C. (1978), *Golden Age of Buses,* Routledge & Kegan Paul, London.
Menzler, F. (1951), 'Lord Ashfield,' *Public Administration,* Vol XXIX,
   London.
Morrison, H. (1933), *Socialization and Transport: The Organization of
   Socialized Industries with particular reference to the London Passenger
   Transport Board,* Constable, London.

Newman, S. 'The trolleybus in London' in Morris, O. (1953), *Fares Please: the Story of London's Road Transport*, Ian Allan, London.

Pearce, M. and Stewart, G. (1992), *British Political History 1867-1990: Democracy and Decline*, Routledge, London.

Pick, F. (1936), 'The Organization of Transport with Special Reference to the London Passenger Transport Board', *Journal of the Royal Society of Arts* Vol LXXXIV, London.

Thomas, D. (1970), *London's Green Belt,* Faber, London.

Urwick, L. and Brech, E. (1953), *The Making of Scientific Management Volume 2 - Management in British Industry*, Pitman, London.

Warren, K. (1980), *Fifty Years of the Green Line,* Ian Allan, London.

Wiseman, R. (1986), *Classic Tramcars*, Ian Allan, London.

3 The origins and development of the passenger transport areas

Kevin Hey

Introduction

This chapter traces the development of passenger transport areas in England as a response to public transport issues facing urban areas. The rationale and process for establishing the initial areas in the major conurbations of Merseyside, South East Lancashire, North East Cheshire (i.e. SELNEC), Tyneside and the West Midlands are considered in the context of earlier developments in local government. The postwar process to reform the structure of local government whilst lengthy and complex illustrates the extent to which public transport issues have been influenced by concerns other than transport. Although these attempts at reform produced only limited adjustments they did provide the identification of a nucleus of local authorities which were part of the conurbations and some of these formed the basis for the initial passenger transport areas. The extensive reorganization of local government which followed in 1974 involved the alteration of these areas along with the creation of two additional transport areas covering South and West Yorkshire. The impact of these changes demonstrates that the current delineation of the passenger transport areas are primarily a product of considerations of local government rather than those of transport. Despite further reforms to local government in the conurbations introduced in 1986 and a wider debate about the structure and role of local government in general the transport area model has proved remarkably robust in practice. Yet the passenger transport areas are not immune to change. The exclusion of significant parts of the hinterlands from the areas has been compounded by the continued expansion of the commuter belt. This has weakened the definition of the transport areas still further. The chapter concludes by demonstrating that they remain trapped by the boundaries and fortunes of local government with little prospect of change.

The founding ancestors

For most of this century responsibility for transportation in the conurbations was spread amongst a collection of different local authorities. Similarly, local public transport services were provided by a number of undertakings of varying size and type with municipal bus operators confined mainly within their respective local authority boundary. Railway services were latterly under public ownership and control. Over the years there were a number of attempts by local authorities in some of these areas to establish joint transport arrangements covering a number of localities. These efforts received fresh impetus after the creation of the London Passenger Transport Board (LPTB) in 1933 (see Chapter 2) which, with the exception of the major railway companies serving the metropolis, brought public transport services under the responsibility of a single public corporation. However in the provinces all attempts to secure some form of voluntary joint public transport body at conurbation level invariably faltered through a combination of municipal pride and suspicions of the intentions of others (Hibbs, 1968, pp.195-97).

The continuation of urban development and expansion coupled with changes to modes of travel brought radical alterations to travel patterns. The effect of these changes made agreement between various units of local government and public transport operators at both policy and operational levels essential for dealing with transportation and public transport issues. Such co-operation and co-ordination as was achieved would often be characterized by a long and laborious process, on occasions with considerable rancour and discord between and within local government units. In these circumstances it is hardly surprizing to find the requirements of the travelling public sometimes relegated as a result of the inevitable political and operational compromises. With transport problems continuing to deteriorate the need for a different structure for dealing with them became essential. It was apparent that voluntary arrangements alone would not produce suitable structures and some form of government intervention would be necessary.

A process of evolution

During the postwar period there was growing acceptance that the structure of local government created in the latter part of the nineteenth century was obsolete and no longer reflected the social and economic realities of modern life (Richards, 1980, pp.33-35). However there was no consensus about the most suitable way to proceed or an appropriate structure.

   The first attempt at reform was made under the Local Government (Boundary Commission) Act, 1945, which established the Local Government Boundary Commission. It was dissolved in 1949 having failed to secure a single change. A number of years were to pass before a further serious attempt was made at reform. The Local Government Act, 1958, provided for the establishment of a Local Government Commission for England. The Commission was given the remit of reviewing and proposing changes to the organization of local government, although in general the scope of the proposals which it could make were quite restricted.

   The most significant innovation was the treatment of the major conurbations. These were recognized as being a distinct local government issue and designated as Special Review Areas (SRAs) (Pearce, 1980, pp.72-73). Five conurbations were identified: Merseyside, South East Lancashire, Tyneside, West Midlands and West Yorkshire. Each area was based upon the principal city and adjacent urban area: Merseyside covered Liverpool, Birkenhead, Bootle, Wallasey and 10 other authorities; South East Lancashire was centred on Manchester with Bolton, Bury, Oldham, Rochdale, Salford, Stockport and no fewer than 46 additional local authorities; Tyneside was based upon Newcastle followed by Gateshead, South Shields, Tynemouth and 11 other localities; the nucleus of the West Midlands was Birmingham with Dudley, Smethwick, Walsall, West Bromwich, Wolverhampton and 20 other council areas forming the urban mass; and finally West Yorkshire which incorporated Bradford, Dewsbury, Halifax, Huddersfield, Leeds, Wakefield and 32 additional localities.

   The Commission received wide powers to determine the most appropriate structure of local government in these areas. The early recommendations which it made were rather cautious but it later became more radical (Wood, 1976, pp.15-20). The first proposals covering a SRA were presented in 1961 for the West Midlands. A structure based upon a number of large unitary authorities was advocated although the complex process to secure implementation meant that it did not take effect until 1966. In 1963 a different solution was proposed for Tyneside with a recommendation of a two-tier structure involving a continuous county and four district authorities. The only other SRA to be examined in detail was West Yorkshire. The result was particularly notable for in 1964 the Commission came to the view that the constituent cities and towns were not a continuous urban conglomeration. There were many parts of the area which were composed of open country and rural in character. This marked West Yorkshire as quite distinct and separate from the other SRAs. It was to prove an important factor some years later when consideration was given to the possible areas which would intially be granted passenger transport area status.

The territory covered by the SRAs was subject to some adjustment. During 1961 the Tyneside area was extended to cover Boldon, parts of Castle Ward, Chester-le-Street, Seaton Valley and Washington. Proposals were also presented for extending the two North West review areas. These represented a more fundamental and major alteration since it was possible to argue that the locations to be included were not wholly urban extensions of the conurbation (Wood, 1976, pp.17-18). These proposals were considered by the Conservative Government but the 1964 general election prevented any further action. The return of a Labour Government committed to a reforming agenda was followed by a series of developments covering local government and transport. The appointment of Richard Crossman as Minister of Health and Local Government brought new vigour to the drive for structural reform. Initially he attempted to utilize the established arrangements but they were complex and time consuming. In 1965 a limited extension to the SRAs in the North West was sanctioned. Merseyside gained Formby, Prescot, Runcorn, Widnes and part of Whiston, whilst South East Lancashire included part of Chapel-en-le-Firth, Glossop, Longdendale, New Mills, Ramsbottom, Saddleworth, Tintwistle, Turton, and Whaley Bridge.

Despite these adjustments it appears that Crossman became increasingly disillusioned with the process of reform (Wood, 1976, pp.15-20). He became convinced that a thorough examination of the system was necessary and in September 1965 he announced proposals to establish a committee to investigate and determine the most appropriate structure for local government (Pearce, 1980, p.110). The Local Government Commission was dissolved having put forward final proposals in only 3 of the 5 SRAs. It had secured only limited achievements largely as a result of the inadequate terms of reference and the extensive consultation process with which it was required to conform (Byrne, 1992, pp.32-34). There was little doubt that the structure of local government, especially in the major conurbations, needed a radical overhaul and that this could only be achieved by direct action from central government. From a transport perspective the real significance of the attempts at reform was not only that any reorganization of local government in the conurbations would also bring structural changes to public transport but that the areas to be covered by such reform were already identified by reference to specified local authorities. The most notable feature arising from the examinations of these areas was the clear distinction between West Yorkshire and the other conurbations. It would be hard to sustain an argument for treating West Yorkshire in the same way as the conurbations of Merseyside, South East Lancashire, Tyneside and West Midlands which demonstrated a closer degree of similarity in terms of each forming a continuous urban mass.

Early in 1966 a Royal Commission on Local Government in England was established under the chairmanship of Lord Redcliffe-Maud. It was charged with the task of producing a reformed structure of local government. Whilst these developments were taking place Barbara Castle had been appointed as Minister of Transport in December 1965 with the remit of producing the integrated transport policy to which the Labour Government was committed (Castle, 1990, pp.37-40). She faced a gargantuan task. The transport difficulties in the conurbations were just one element of a number of issues facing the new Minister. In July 1966 she was able to present some guiding principles in the White Paper 'Transport Policy' (Ministry of Transport, 1966).

The framework for urban transport was based upon the premise that the transport problems in the major conurbations were so acute that reform could not await the general reorganization of local government which was likely to be some years away. For these areas it was proposed to establish conurbation transport authorities which would have responsibility for public transport over an urban region. They were to be an adhoc interim measure pending the findings of the Royal Commission on Local Government whereupon they could be readily absorbed into single local authorities covering an entire urban region which would have responsibility for transportation also.

The proposals for conurbation transport authorities met with a mixed response but the Minister was not to be deterred from developing the concept further. As part of this process she toured the United States of America during October 1966 to examine developments in passenger transport. In December 1967 the government published a further White Paper entitled 'Public Transport and Traffic' (Ministry of Transport, 1967) in which the proposal for conurbation transport authorities was replaced by the concept of passenger transport areas. The change of name was more than semantic since transport areas could be established in any area of Britain and would not necessarily be restricted to the conurbations. It was widely believed that these revised proposals were modelled to a large extent on the Massachusetts Bay Transportation Authority which the Minister had observed during her visit to the United States. Local public transport was to be retained as a local government function through which integrated transport plans could be developed. Accordingly it was proposed to establish in these areas a Passenger Transport Authority (PTA) with wide ranging responsibilities for overall transport policy. The PTA would consist primarily of councillors drawn from various local authorities in the area together with nominees from the Ministry of Transport. A professional body in the form of a Passenger Transport Executive (PTE) would assume responsibility for the planning, development and provision of public transport services.

The delineation of passenger transport areas was to adhere to the concept of establishing areas which had coherence in transport terms. This meant areas far larger than any of the local authorities then in existence. The freedom to designate an area without restraint or contamination from existing administrative boundaries, which in any event were acknowledged as no longer having relevance to transport requirements, was more illusory than real. Nevertheless for the first time the possibility was presented of drawing the map based primarily upon transport requirements.

The government added two guiding principles (Ministry of Transport, 1967, para.8-19). The first of these was that passenger transport areas should include the places of substantial commuter hinterland. There was growing acceptance that the continuing expansion of urban areas and the changing patterns of social life had increased the degree of interdependence between town and country. The inclusion of places of substantial commuter interest within passenger transport areas was a logical step. The second principle was that passenger transport areas should correspond with those of existing or proposed local authorities. Unlike the first principle, which was complementary to the desire of defining areas which had coherence in transport terms, the second principle sat rather uneasily with this concept. It was clear that on an individual basis the boundaries of each local authority no longer reflected transport realities. It was possible to argue that a number of adjoining authorities could produce an area which had greater coherence in transport terms and that using local authority boundaries had a certain logic, not least in terms of simplicity by virtue of linking transport and planning areas. The main problem was that the arrangement continued to subjugate transport to local authority boundaries which were clearly determined by a variety of considerations other than transport.

Designating the passenger transport areas

The White Paper 'Transport Policy' had provided only an outline of the concept of transport areas for the conurbations. It gave little indication of which places would receive such status but the most likely starting point were the five SRAs under the previous examination of local government. During the autumn of 1966 Barbara Castle visited the largest local authorities and transport operators in all of the SRAs, although of these areas only South East Lancashire, Merseyside, Tyneside and the West Midlands had been confirmed as conurbations by the earlier Local Government Commission. This provided the real clue to the likely passenger transport areas and these four were duly confirmed as the intended transport areas in a parliamentary written answer on 10 May 1967 (Official Report, 1966-67, col.236). Despite numerous adverse comments from many of the local authorities likely to be affected by the proposals the Minister was not persuaded to alter her decision.

The position of the remaining SRA which covered West Yorkshire was more interesting. In view of the conclusions of the Local Government Commission some years earlier that the area did not form a continuous urban mass it was probably rather difficult for the Ministry of Transport to justify treating West Yorkshire on the same basis as the other SRAs. Although transport areas could be established in any area it seems likely that the Ministry of Transport was content at this stage for them to adhere to the original concept based upon conurbations and as a result West Yorkshire was excluded from the initial areas.

Proposals to put passenger transport areas into effect formed part of the Transport Bill presented to parliament on 7 December 1967. Even though the Bill was still in the parliamentary process the Ministry of Transport engaged in confidential and informal consultations with local authorities in the proposed transport areas during April 1968. The consultations were based on two listings for each area. The first list consisted of those localities which were part of the continuous urban belt and where the balance of advantage favoured inclusion in the transport area. By contrast the second list was composed of those localities with sufficiently important passenger transport links with the conurbation for inclusion to be considered. The places in the primary list included all the major authorities which were within the SRAs. By contrast the second list covered a combination of authorities within and adjacent to the SRAs. Merseyside could have covered Runcorn and Widnes which were part of the SRA along with Southport which was not; for SELNEC the places in the former category were Glossop, Horwich, Ramsbottom, Saddleworth, Tintwistle, Turton and Westhoughton whilst Atherton and Leigh had been excluded from the SRA; the Tyneside localities of Boldon, Seaton Valley and Washington were part of the extended SRA with Prudhoe and Sunderland providing examples of adjacent authorities; whilst the West Midlands could extend to Bromsgrove, Cannock, Redditch, Rugeley and Tamworth. The matter of inclusion or exclusion from the passenger transport areas was not the only issue of concern as the level of representation by the local authorities upon the PTA was a particularly contentious issue in view of the balance of power between the various constituent authorities.

Once the Transport Act, 1968, received Royal Assent these unofficial approaches were followed by the formal consultation process. Final details of the initial passenger transport areas were laid before parliament on 7 February

1969. The main criterion for determining which local authorities were to be included in the areas was the travel-to-work information provided by the 1966 census. This produced boundaries which were an amalgam of the original and extended SRAs. There were some differences: Merseyside excluded the areas of Ellesmere Port, Runcorn and Widnes; SELNEC incorporated Atherton, Leigh and Tyldsley but did not cover New Mills or Whaley Bridge; Tyneside failed to secure Boldon and Washington; whilst the West Midlands also covered Bromsgrove, part of Lichfield and Redditch. The places excluded from the transport areas were those on the periphery where there were differences of opinion about the desirability of the authority forming part of the area. Clearly some of these localities had economic and social links with the conurbation but since the transport areas were an interim arrangement pending the outcome of the review of local government the Ministry of Transport preferred to consider extending them at a later date upon local government reorganization.

The Redcliffe-Maud report

The Royal Commission on Local Government in England reported in 1969. It had identified a number of problems with the structure of local government then in existence. One of the major deficiencies stemmed from the structural distinction between town and country which was not only at variance to the realities of social and economic life but also hampered the ability of local government to meet the challenges which the Commission believed lay ahead (Redcliffe-Maud, 1969, Vol.I, para.6-9). The Royal Commission concluded that the structure of local government needed changing. There was broad agreement on the principles which should provide the basis of the new structure but beyond this there was a clear difference in approach between one member of the Commission and the majority view.

The proposals advocated in the main report of the Commission was for a unitary structure of local government except in the three conurbations of SELNEC, Merseyside and the West Midlands. In these areas a two-tier structure incorporating a metropolitan form of government at conurbation level was proposed. It should be noted that a unitary structure was proposed for the Tyneside and West Yorkshire areas rather than the metropolitan arrangement. The recommendation of a form of metropolitan government was a departure from previous principles. It followed closely the model advocated by the Royal Commission on Local Government in Greater London 1957-60 under the chairmanship of Sir Edwin Herbert which resulted in the creation of the Greater London Council (GLC) and 32 London Boroughs in 1965.

The metropolitan authority would have responsibility for strategic issues, including transportation and public transport, whilst a number of smaller district authorities would provide the personal services. It was recognized that strategic planning, transportation and public transport issues could only be dealt with effectively under one authority and that this should be at the city region level in order to conform to a spatial pattern which matched the catchment area of the transport problem.

The Commission was distinctly conservative about the boundaries of the new units of local government since these were to be formed from existing authorities. This approach was advocated on the grounds of maintaining the traditions, loyalties and momentum of the existing system. Indeed boundaries were only altered where this was supported conclusively by the evidence for such a change (Redcliffe-Maud, 1969, Vol.I, para.279-281). In all cases the area of the respective metropolitan (or unitary) authority was more extensive than the passenger transport areas. As the Commission observed this was to be expected given the wider remit of the review of local government where the focus was upon the creation of administrative units for the conurbations which incorporated the surrounding areas with which they had strong social and economic links (Redcliffe-Maud, 1969, Vol.I, para.328-330). This certainly provided some support for the view that the original passenger transport areas were too small.

A minority report developed by Derek Senior accepted the deficiencies of the existing system but advocated an alternative two-tier structure based upon the city region principle for England as a whole. Although the proposals were well received in those academic circles which favoured an approach based on social-geography (Pearce, 1980, pp.121-22) the model attracted little in the way of general support. From the point of view of the passenger transport areas then in existence it would have resulted in considerable expansion of the territory covered by them. (Redcliffe-Maud, 1969, Vol.II, para.379).

The ten years from 1960 to 1969 showed vividly the problems associated with attempts to reform the local government structure, especially in the conurbations. At the heart of the controversy were two issues. The first stemmed from the drawing of the metropolitan area boundaries which was fraught with problems. The second issue focused upon the most appropriate structure and here debate raged between advocates of the unitary and two-tier systems. The variety of proposals advanced for some of the conurbations provided the clearest evidence of these conflicts. For example at various stages the Tyneside area had been recommended both for a unitary and two-tier structure; the review of the West Midlands was followed by a reorganization on a unitary basis but the area was now recommended for a two-tier structure; whilst examination of West Yorkshire had produced in each case a different conclusion and solution. For those seeking a coherent approach to local government in the conurbations the situation could not be other than confusing.

Local government reorganization

The Labour Government accepted the majority report of the Royal Commission with few amendments. The most significant change was the increase in the number of metropolitan areas by adding West Yorkshire and South Hampshire. The government commenced consultations with interested bodies on the proposals but the general election during 1970 interrupted progress. Not for the first time was a general election to change the course of events with regard to the structure of local government. The Conservative Party were returned to power and promoted an alternative set of proposals based upon the two-tier principle.

Whilst the extent to which the revised proposals were consistent with those of the Royal Commission was open to debate (Hampton, 1991, p.36) they were subject to some modification prior to coming into effect in 1974. Although the city region concept for the conurbations was retained the metropolitan counties which were created differed in at least three respects from the recommendations of the Royal Commission: first, the number of such counties was doubled with the designation of additional areas covering South Yorkshire, West Yorkshire and Tyne and Wear; second, the boundaries of these areas were drawn much more tightly around the conurbations compared to the initial proposals; third, the authorities were given fewer functions than originally envisaged although they did retain responsibility for strategic planning, transportation and public transport.

There were many criticisms of the reorganized system (Byrne, 1992, pp.44-45) with the boundaries of the metropolitan counties proving a particular source of controversy. It was argued that drawing the boundaries so close to the urban areas produced authorities which were too small as they failed to incorporate sufficient of the rural hinterland and perpetuated the institutional distinction between town and country (Jones, 1973, p.160). The explanation for such an outcome is not hard to find. The creation of metropolitan authorities presented a major threat to nearby shire counties such as Cheshire, Lancashire, Staffordshire and Worcestershire. The Conservative Government was sympathetic to retaining the shire counties as it had considerable political strength within them and it was prepared to accept boundary changes as long as the basic principles of the reorganization could be maintained (Hampton, 1991, pp.35-37). These two factors ensured that the proposals for local government reorganization were to be subject to considerable change.

Not unnaturally the shire counties adjacent to the metropolitan areas sought to preserve the maximum sphere of influence by retaining as much as possible of the traditional county boundary. These campaigns struck a chord in those places on the border between the proposed metropolitan and the traditional shire counties where in many instances the local populace mounted a spirited defence to avoid becoming part of a metropolitan area. Whilst the motivation for such campaigns was complex there was a strong desire by the suburban middle classes to retain the association of affluence which the hinterland enjoyed as a result of location within a shire county notwithstanding the opportunity to continue avoiding city rate contributions (Kingdom, 1991, p.82).

For the passenger transport areas the effects of local government reorganization were complex. The new arrangement did at least bring all the initial passenger transport areas under a metropolitan county authority which also became the PTA for the area, whilst the desire to ensure a uniform structure over the metropolitan areas resulted in the creation of two new passenger transport areas covering South and West Yorkshire.

Of the existing areas SELNEC was retitled Greater Manchester whilst Tyneside became Tyne and Wear to better reflect the localities covered. The boundary of each passenger transport area was made to correspond with the respective metropolitan county which resulted in an increase to the total size of the transport areas as well as the removal of a number of places from the initial areas. On the positive side the SELNEC area was altered to include Wigan; Merseyside gained St.Helens and Southport which during the creation of the original transport area had been identified as having sufficient links with the conurbation to warrant inclusion; the same situation applied with regard to Sunderland which represented a major extension to the original Tyneside area; and Coventry joined the West Midlands. The places removed from the original areas varied from those which were part of the conurbation to those which although on the periphery had significant social and economic links with the area. For example in Merseyside parts of West Lancashire and Whiston along with Neston were removed from the passenger transport area whereas Greater Manchester excluded Alderley Edge, Disley, Glossop, Tintwistle, parts of Turton, Whitworth and Wilmslow. On Tyneside parts of Castle Ward, Seaton Valley and Whitley Bay which were within the original passenger transport area were outside the metropolitan county. In view of the strong links which these places had with the Tyneside conurbation the Passenger Transport Authority and Executive were moved to comment that this 'serves to illustrate that it is difficult to decide the most appropriate boundary and that administrative boundaries do not necessarily coincide with logical boundaries of transport operation' (Tyneside PTE, 1973, p.12). The West Midlands transport area no longer covered Bromsgrove, Cannock, Redditch and parts of Seisdon and Meriden.

On balance the verdict of local government reorganization upon the passenger transport areas was that the metropolitan counties were a step in the right direction. The transport areas were fully integrated into the structure of local government as originally envisaged. Additional urban localities were brought under the transport area arrangement and although some places were removed from the original areas these were not significant by comparison with the expansion which had occurred. There were also adverse consequences. The passenger transport areas were now exclusive to the conurbations as a feature of metropolitan local government and as such this was a return to the original concept of conurbation transport areas. The other consequence was that the areas suffered through the general erosion of the metropolitan counties during the reform process with the subsequent retention of Skelmersdale in Lancashire at the expense of Merseyside whilst Greater Manchester failed to secure Poynton. There were other notable omissions - Ellesmere Port, Runcorn and Widnes remained outside Merseyside; Skipton and Harrogate were excluded from West Yorkshire; and Warrington escaped inclusion in either Greater Manchester or Merseyside (Redcliffe-Maud and Wood, 1974, pp.46-49). The end result for the initial passenger transport areas was that they by no means covered the continuous urban area of each conurbation whilst those in Yorkshire had significant areas which were rural in character. The architects of the reorganized system had hoped to end for a generation the uncertainty which had surrounded local government during the postwar period (Alexander, 1982, p.10). It was a worthy aim but the hope was to prove misplaced.

Streamling the cities

Barely a decade later the structure of local government in the conurbations was subject to further reform. The metropolitan authorities were latterly under Labour control and a number of them adopted polices which were at variance with the objectives of the Conservative Government. Public transport subsidy was one aspect which brought these authorities into conflict with central government. In the 1983 general election the Conservative Government was re-elected with a manifesto committment to abolish the metropolitan county councils. On return to office these proposals were presented in the White Paper 'Streamlining the Cities' (Department of Environment, 1983). The general assessment by the government was that the conurbation level authorities had too few functions, found difficulty in establishing a strategic role, and had persued policies which conflicted with those of national government. The case for abolition was far from convincing which has led to the suggestion that the primary motive for the reorganization was due to the fact that the authorities were Labour controlled and had become increasingly critical of central government (Chandler, 1991, p.27).

After lengthy and contentious debate the Local Government Act, 1985, abolished the metropolitan county councils in 1986. It is significant that the key functions of these authorities, including passenger transport, were maintained at conurbation level and transferred to statutory joint boards. The passenger transport area concept was retained and the PTAs reverted to joint bodies similar to those which applied to the initial areas prior to 1974. The remaining functions of the metropolitan counties were either distributed to the district councils or made subject to voluntary joint arrangements. In most of the metropolitan areas the district authorities developed some form of voluntary joint body covering varying aspects of the transportation function.

The myriad of joint arrangements which now exist have led some observers to claim that a conurbation level of local government has been retained albeit in a different form and with redefined political status (Leach, Davis, Game and Skelcher, 1990, pp.3-4). It could be argued that retention of the passenger transport areas provided confirmation of the concept but the reality was probably more mundane. The placing of public transport responsibilities along with other major functions under statutory joint arrangements was an appealing option for a central government committed and anxious to secure the swift abolition of the metropolitan county authorities. Nevertheless in the conurbations it is clear that passenger transport areas are a more suitable unit for addressing transport problems than fragmenting responsibility amongst a variety of authorities. Indeed despite the various reorganizations of local government during the last 25 years, which have included the rise and fall of the city region concept, the passenger transport areas have demonstrated remarkable resilience. In this period of rapid change the PTA arrangement has shown that it is both robust and adaptable with the capability to be organized around either a unitary or two-tier structure.

Current issues

Any claims that the passenger transport areas have to coherence in transport terms must now be open to question. The initial exclusion of significant parts of the commuter hinterland was a serious omission which has been compounded by the continued expansion of the commuter belt. In view of these developments and the continuing debate about the structure of local government the time would now seem appropriate for considering the case for re-examining the passenger transport areas.

The social and economic environment in which the passenger transport areas were created has changed considerably. It is difficult if not impossible to point to one particular change as the most significant since there are a number and the interactions between them have all contributed in varying degree to the current situation. During the postwar period a number of new demographic trends have emerged. Britain entered a period of counter-urbanization in which the conurbations experienced both a relative and absolute decline in population. A combination of planned redistribution schemes and voluntary migration account for this change. All areas suffered population loss although the rate of decline has been far from uniform. Some of this loss has been to the outer suburbs and into the expanding commuter hinterlands. A combination of an improved road network, increased car ownership and the allure of small town and rural life have all played a part towards increased commuting. Mass commuting over longer distances has strengthened and made ever more complex the travel patterns between the conurbations and hinterlands.

The transport environment has changed in other ways too. A combination of privatization and deregulation has replaced social ownership of public transport operators along with the removal of centralized planning and strict quantity control. Local government still retains a legitimate though much reduced role in local passenger transport and it certainly could be argued that in such an environment the concept of a passenger transport area is a relic from a different age. Such an argument might be sustainable had the transport problems in the conurbations not continued to deteriorate but the reality is that the difficulties of urban transport congestion are now more widespread, more acute and more protracted than ever. In these circumstances, the concept of transport areas at conurbation level defined by reference to commuting and other social travel patterns remains persuasive.

The inextricable link between the boundaries of the metropolitan areas and those of passenger transport areas needs to be addressed. The crucial issue is the proposition that the two areas should be conterminous particularly since the boundaries of current local government areas were the product of the political process rather than one which attempted to reflect the pattern of social and economic life. This must surely bring into doubt the validity of the current metropolitan government model as a basis for delimiting passenger transport areas. The areas are certainly a closer reflection of local government considerations rather than those of transport. This link has other implications. It effectively condemns the passenger transport area boundaries to a relative permanence. After their creation nearly a quarter of a century ago the atrophy of these areas is not entirely unexpected. A clue to this outcome can be found in the relative durability of local authority boundaries since the end of the nineteenth century and the limited effectiveness of the machinery for reviewing them during the intervening period (Stanyer, 1976, p.73). The changes made were often subject to a protracted and hostile process between competing local authorities engaged in maintaining existing boundaries regardless of the changing nature of the pattern of life.

The Local Government Boundary Commission established under the Local Government Act, 1972, to review local authority areas had power to designate additional areas as metropolitan authorities. The potential effect of any such proposal on an area was significant as it would have necessitated a reallocation of local government functions on a metropolitan basis including passenger transport area status. This provided an opportunity for redressing the restrictive boundaries of the metropolitan counties along with those of the passenger transport areas but in the event the Boundary Commission never ventured such a change.

The Commission was dissolved in 1992 and replaced by the Local Government Commission for England which now has the task of reviewing the structure of local government. The work of this Commission has concentrated on reviewing the shire counties under policy guidance from central government which has demonstrated a clear preference for unitary authorities. As with the abolition of the metropolitan county authorities a decade earlier it has been argued that the case for a unitary structure was assumed rather than made (Stewart, 1994, pp.13-15). Although the work of the Commission has focused upon the shire counties it would be a mistake to believe that this will have little impact on the metropolitan counties. The reviews have in fact confirmed the present boundaries of the metropolitan areas and hence the passenger transport areas by virtue of maintaining the existing boundaries of the shire counties. As a result the defects of the restrictive boundaries of the metropolitan areas will remain and with them the boundaries of the passenger transport areas will be consigned to a further period of stagnation. This might be the lesser of two evils for within the review of local government there is a potentially more serious threat. Although there is no restriction on boundary changes between metropolitan and shire counties section 14(7) of the Local Government Act, 1992, provides for new areas created on this basis to have shire rather than metropolitan status. The effect of such a change upon a passenger transport area is far from clear since there are a number of options. One alternative is that the passenger transport area might remain unchanged albeit incorporating part of an authority in a shire county, on the other hand the new authority could be removed from the passenger transport area altogether.

Bearing in mind the link between passenger transport areas and local government the real issue is to identify which structural arrangement for local government might be most suitable from a passenger transport area perspective. Although there are a number of alternative structures for metropolitan government the notion of the city region remains the most suitable for conterminous passenger transport areas. Whilst adoption of this concept in the current review of local government is rather remote it is worth considering the impact which such a model might have upon the transport areas. The existing metropolitan areas could be extended with, for example, Ellesmere Port, Widnes and Runcorn incorporated into Merseyside; Wilmslow and Knutsford would join Greater Manchester; whilst Redditch and Tamworth would be included in the West Midlands (Leach, Davis, Game and Skelcher, 1990, pp.269-74). As a consequence the existing passenger transport areas could be extended to cover an area far larger than the current definition so removing the worst anomalies of the present boundaries. They would then more accurately reflect the realities of social and economic life particularly in regard to commuting patterns and have a greater claim to coherence in transport terms by virtue of definition at a spatial level which corresponds with the catchment area of the transport problem.

This surely is the rub for areas covering the conurbations and hinterland which make sense in terms of dealing with public transport do not necessarily make for units of local government which find ready and popular acceptance. Perhaps the real message is that in the absence of a city region model of local government the time has come to consider the definition of passenger transport areas detached from strict adherence to the existing metropolitan areas and the consequences of the inevitable political struggles which accompany the search for an effective and convenient form of local government.

Conclusions

The passenger transport areas created in 1968 were an adhoc arrangement pending local government reorganization. The boundaries of these areas followed, with minor variations, those of the local government SRAs identified in 1958. The subsequent development of the transport areas was determined by considerations of local government which were not necessarily the most appropriate from a transport perspective. In the meantime the social and economic environment has changed markedly and this has further weakened any claims which the areas had to coherence in transport terms. There is once again need for an informed debate about the arrangements for public transport in these areas. With the transport problem now acute and continuing to deteriorate the current definition of the transport areas and structural arrangements for metropolitan government do not seem particularly well suited for dealing with it. The boundaries of the transport areas are in need of review along with consideration of the most suitable institutional framework to support the growing consensus that public transport should once again play a greater role in urban transport. For too long the issues of public transport have been relegated in the debate about local government by other considerations. It is time that the topic of public transport was promoted to the top of the agenda - the worsening transport conditions in our major conurbations would seem to demand no less.

References

Alexander, A. (1982), *Local Government in Britain since Reorganization*, Allen & Unwin, London.

Byrne, T. (1992), *Local Government In Britain*, Penguin, Harmondsworth.

Castle, B. (1990), *The Castle Diaries 1964-76*, Papermac, London.

Chandler, J. A. (1991), *Local Government Today*, Manchester University Press, Manchester.

Environment, Department of (1983), *Streamlining the Cities*, HMSO, London.

Hampton, W. (1991), *Local Government and Urban Politics*, Longman, Harlow.

Hibbs, J. (1968), *The History of British Bus Services*, David & Charles, Newton Abbot.

Jones, G. (1973), 'The Local Government Act 1972 And The Redcliffe-Maud Commission' in *The Political Quarterly*, Volume 44, Political Quarterly, London.

Kingdom, J. (1991), *Local Government and Politics in Britain*, Philip Allan, Hemel Hempstead.

Leach, S. Davis, H. Game, C. and Skelcher, C. (1990), *After Abolition: The Operation Of The Post-1986 Metropolitan Government System In England*, Institute of Local Government Studies, University of Birmingham, Birmingham.

Official Report, Fifth Series, (1966-67), *Parliamentary Debates*, Volume 746, London.

Pearce, C. (1980), *The Machinery of Change in Local Government 1888-1974*, Allen & Unwin, London.

PTE, Tyneside (1973), *Public Transport On Tyneside: A Plan For The People*, Tyneside Passenger Transport Authority, Newcastle upon Tyne.

Redcliffe-Maud, Lord (Chairman) (1969), *Report of the Royal Commission on Local Government in England 1966-69*, Cmnd.4040, HMSO, London.

Redcliffe-Maud, Lord and Wood, B. (1974), *English Local Government Reformed*, Oxford University Press, London.

Richards, P. G. (1980), *The Reformed Local Government System*, Allen & Unwin, London.

Stanyer, J. (1976), *Understanding Local Government*, Martin Robertson, Oxford.

Stewart, J. (1994), 'The Flawed Process' in Leach, S. (ed.) *The Local*

Government Review: Key Issues and Choices, Institute of Local Government Studies, University of Birmingham, Birmingham.

Transport, Ministry of (1966), *Transport Policy*, Cmnd.3057, HMSO, London.

Transport, Ministry of (1967), *Public Transport and Traffic*, Cmnd.3481, HMSO, London.

Wood, B. (1976), *The Process Of Local Government Reform 1966-74*, Allen & Unwin, London.

4 Municipality and civic pride in Canada: Toronto 1895-1995

Stephen Shaw

Introduction

In the late nineteenth century, the rapid and uncontrolled growth of many North American cities created squalid, and potentially volatile social conditions, as successive waves of immigrants arrived seeking work and accommodation. In general, their central business districts lacked the imposing architecture, monuments and public open space of their counterparts in the Old World. And, despite their 'gridiron and block' street patterns, traffic circulation was becoming chaotic and slow. By 1900, however, an influential group of civic-minded and visionary intellectuals were promoting the concept of the 'City Beautiful'. With an emphasis on unity and elegance of civic design, the American metropolis could be demolished and sublimely transformed into a proud, orderly, and civilized commercial-administrative centre. The improvement of transport infrastructure would be an important feature of the grand design. Despite much lofty rhetoric, little of substance had been achieved on the ground by the start of the World War One. Nevertheless, in the Dominion of Canada, certain of the American City Beautiful movement's principles became fused with municipal reformism, and the ideals of the early town planning movement from Britain. Thus, a belief in civic improvement and public service continued to exert a strong influence.

This chapter explores the reasons why a number of leading municipalities in Canada developed an ethos of civic pride which embraced a far greater level of interventionism than their counterparts in the United States. A study of Toronto's public transport system over the past century is used to illustrate the acceptance of municipal ownership, and to examine the justification for active involvement of city government in mass transit and other local utilities. The study describes the unification, modernization, and expansion of the electric streetcar network as a public service. Notwithstanding the huge rise in car ownership, and decentralization of population in the postwar era, the city's public transport system has continued to play a critical role in facilitating economic growth, especially after the urban expressway programme was abandoned in the early 1970s. With an explicit goal of creating an attractive environment for inward investment, there was remarkably broad based political support for public expenditure on commuter railways, subway and rapid transit lines, along with other prestigious amenities to establish Toronto as a 'world city'.

Thus, throughout the past century, public transit and other services regarded as important civic amentities, have attained a high profile in the local politics of urban Canada. Over the past 25 years or so, there has also been a new wave of radical reformism - one which has challenged the civic and business establishment with a concern for social justice, and demands for citizen participation in municipal services. The final section of the study examines the influence of urban social movements which have lobbied city government to set a new agenda, with an emphasis on sensitivity to the needs of disadvantaged neighbourhoods and communities. As greater recognition is given to the rights of disabled people, women, and ethno-racial minorities as service users, the concept of 'civic pride' is currently being re-examined and reinterpreted by municipal service providers, with recognition of the need to overcome the various barriers - physical, social, and linguistic/cultural which particular groups in the local population experience in making full use of public services.

Cities fit for merchant princes

At the turn of the twentieth century, the ideal of the City Beautiful seems to have fired the imagination of civic improvement societies and lobby groups throughout North America. The 'White City' of the World's Colombian Exhibition, held in Chicago in 1893, had promoted a powerful aesthetic, linking 'orderliness and beauty'. Daniel Burnham, the architect appointed director of the World Fair, wanted to demonstrate that the American city of the coming century could be rebuilt and refashioned on a grand scale. Its neoclassical splendour would emulate the achievements of mercantile capitalism in

Renaissance Italy. Monumental civic buildings would provide the centrepiece. These would be admired from broad, tree-shaded streets and plazas, with landscaped public parks for healthy outdoor recreation. Transport improvements would be a key element of the city plan featuring:

*   imposing boulevards and pleasure drives;

*   magnificent Union stations and trans-shipment centres for freight;

*   use of the new technology of electric traction for streetcars, subway lines and elevated railways.

The White City was merely an elaborate stage set, demolished when the exposition closed, but the movement which it inspired promised to create a new sense of order and permanence. Its aims went far beyond environmental and aesthetic improvement. Urban society itself would achieve a new unity of purpose. C. M. Robinson (1913, p.211), one of its chief advocates, went on to develop a belief that the moral and spiritual standards of the people would be advanced by this art, and that their political ideals would 'rise with a civic pride and community spirit born of an appreciation that they are citizens of "no mean city".'

A century after the launch of the City Beautiful, opinion is still divided as to how its underlying ideology should be interpreted. On the one hand, it advocated generosity in the creation of public and semi-public space such as passenger transport facilities. The movement also served to underline the failure of laissez-faire urbanization - its inefficiency and the sheer inconvenience it often created - as well as its physical and spiritual impoverishment. Its ideals provided intellectual guidance for the development of civic and landscape design, land-use controls, and transportation planning. With foresight and planning, these elements could be treated as interrelated elements of a greater whole. Wilson (1989, p.3) has stressed the essential good intent of the movement's protagonists, arguing that it possessed its 'own ideology, purpose, and mode of operation'. Daniel Burnham's impressive Plan for Chicago (1909) encompassed the whole city region within a 60 mile radius of the central business district - the Loop. Although never implemented it is still admired for its proposed creation of an extended lakeside park, boulevards and forest preserves, as well as its reorganization of the city's passenger and freight transport system (Wilson, 1989, p.281).

The City Beautiful's supporters fought long and hard against entrenched opposition from powerful private interests such as railway companies. Their campaigns to remedy the inadequate capacity, inconvenient interchanges and spartan facilities of stations owned by competing railways received enthusiastic popular support.   The campaign for a Union station at Kansas City, for example, spanned two decades of hard negotiation (Wilson, 1989, pp.193-212). In New York, the contract for construction of the first subway (1899) contained a 'Mandate for Beauty'.   The subway was to be owned by the city government, but built and operated on a 50 year lease by a consortium led by financier August Belmont.   As Kiepper (1994, pp.4-5) has commented, from the beginning, there was tension between the public mandate and the profit motive.   'Belmont was a capitalist of the Robber Baron era, not a latter-day Medici prince.   He regarded the subway as a strictly financial venture'.   The stations were conceived as magnificent subterranean palaces, yet Belmont wanted to paper the walls with large advertizing posters to generate revenue and eliminate the need for first-class wall tiling underneath.

On the other hand, the City Beautiful movement's underlying ideology is open to less charitable interpretation. Jacobs (1965) saw the image of the White City as a precursor of the alienating and oppressive landscapes of mid twentieth century modernism.   Marcuse (1980, pp.23-58) discusses its essentially aesthetic, architecturally-oriented view of planning. Yet, the North American bourgeoisie had a dread of popular discord in the wake of the economic recession and dislocations of the 1890s.   Conflagration and riot might all too easily errupt from the neighbourhoods of non English-speaking, recent immigrants. The overarching principles of orderliness and beauty, unity and harmony have sinister connotations in the context of 'Anglo conformity', and social control. Real estate owners around the areas to be upgraded by parks, road and public transport improvements stood to gain as land values appreciated. And, at a macro level, there was a strong element of 'local boosterism': better transport infrastructure and an attractive physical environment would help one city to compete with its rivals for inward investment and enhanced income from local taxes.

The City Beautiful ideal also attracted a great deal of contemporary criticism. The movement received scornful attacks, especially for its lack of concern for the living conditions of inner city dwellers. It was undeniably elitist, and preoccupied with the grandiose. Despite its public-spirited rhetoric and concern for community, it did not appear to offer much material benefit to the city's poorest inhabitants. Through redevelopment on a grand scale, the social character of neighbourhoods could be 'improved', although little was said of the fate of displaced residents.   Housing reformer B. C. Marsh (1908, pp.1514-18) inveighed against its preoccupation with cosmetic display and only rarely could the poor escape from their squalid surroundings 'to view the architectural perfection and to experience the aesthetic delights' of the improvement schemes.   The concept of a city fit for merchant princes had, in any case, become less fashionable by the start of World War One.   Its supporters grew weary and disillusioned with the difficulties of enlisting the co-operation of landowners, developers and utilities such as railway companies.

City governments marvelled at the grandeur of comprehensive and detailed plans for urban reconstruction such as Burnham's Chicago, but balked at the cost of land acquisition and the scale of the proposed public works. In the United States, an alternative planning philosophy - the 'City Practical' - eschewed the notion of civic design on a grand scale in favour of technocratic solutions to specific urban problems which required urgent resolution, notably zoning to separate incompatible land-uses, and management of traffic flows. Advocates of the new functionalist approach, such as George B. Ford, emphasized the relevance of Frederick W. Taylor's ideas on scientific management to public administration. Commercial and civic groups pressed home the need for 'efficient transportation' and looked to municipal officials to devise solutions as cars swarmed on to downtown streets which were poorly suited by width and condition to accommodate them. 'Streetcars fought motor cars for space, and disgorged passengers in the middle of clogged thoroughfares' (Brownell, 1980, p.67).

In Canada, however, the visionary spirit of planning and civic improvement was never extinguished. Plans modelled on the White City ideal had been drawn up for Toronto (1905, 1909 and 1911), Montreal (1906), Winnipeg (1913), Calgary (1914) Ottawa and Hull (1915). As in the United States, few of the proposals contained in these early schemes were implemented due to their enormous cost, and opposition from powerful commercial interests, yet as Artibise and Stelter have shown (1981, pp.20-21), the City Beautiful idea in Canada was far from dead. In the late 1920s, ambitious city plans were commissioned by the municipalities of Vancouver and Toronto. In the aftermath of the Wall Street Crash, the proposals were scaled down. Greber's grand scheme for the federal capital in Ottawa was, however, constructed during the 1930s. Thus, a distinctly Canadian approach to planning evolved, drawing on a range of American, French and British influences, and developing in a more favourable political climate than was the case in the United States.

Local government in Canada, through its colonial origins, shares a legal heritage with municipalities in Britain. Like their contemporaries in Britain, metropolitan authorities such as Toronto acquired powers to carry out a wide range of services for the benefit of local residents and businesses, raising finance through local taxation and borrowing. In addition, some set up commissions accountable to boards appointed by the municipality. Under the British North America Act, 1867, the provinces have jurisdiction over 'local works and undertakings' including urban transport. In practice, however, many of these activities were delegated to municipalities. British ideas and practices were diffused through council members and officers who had recently emigrated from Britain to English-speaking Canada. There were also strong links with Britain through professional bodies and influential individuals. A notable example of the latter was Thomas Adams, a founder member of the British Town Planning Institute, who was appointed advisor to the Canadian Federal Conservation Commission in 1914 (Gunton, 1991, p.95). With great conviction and energy, he publicized the urgent need to tackle social problems comprehensively, arguing that planning should be concerned with 'every aspect of civic life and civic growth' (Adams, 1915, p.149).

Thomas Adam's radicalism and his opposition to 'the deadly doctrine of laissez-faire', was greatly diluted when statutory land-use planning was eventually introduced in Canada in the late 1940s. Nevertheless, the ideal of a comprehensively planned city with attractive and well run services has remained a persistent theme. Canadian municipalities embraced the idea of public ownership and other forms of interventionism with far more enthusiasm than those in the United States. There was a greater willingness to invest in order to improve public transport and other urban amenities. The implications for local taxation and long term borrowing have always been a subject of discussion and the voting power of affluent car owning households in the postwar suburbs has influenced the policy debate on priorities for investment. Nevertheless, in the postwar era, most Canadian civic leaders distanced themselves from what they saw as the 'private affluence and public squalor' characteristic of most American cities. High quality, well managed local services would help attract inward investment, and serve as a catalyst for wealth generation. The following case study explores the influence of these principles of urban government in the development of public transport as municipal enterprize in the city of Toronto and its surrounding region.

Transport for 'Toronto the Good'

Toronto at the turn of the century was experiencing rapid growth. Capital of the province of Ontario, inland port and major transcontinental rail head, it was well located for international trade, and developed a diverse industrial base. The city's commercial and civic leaders formed an exclusive clique of white, Anglo Saxon Protestant men from England, Scotland and Ulster, and over 85% of the workforce and their families were of similar stock. In contrast, many who emigrated to the rival city of Montreal were Roman Catholics from France and Southern Ireland. The pervasive influence of Victorian Protestant morality on the lives of its citizens had earned it the title 'Toronto the Good'. As in Britain, religious fervour and concern for spiritual and social wellbeing were closely associated. William Howland, elected mayor in 1885, led a reform-minded council that waged war on everything they regarded as 'organized sin': the saloon, the gambling den, the brothel, and even the theatre.

Reformers were convinced that 'vice was so much a fact of city life that it menaced the national destiny' (Artibise and Stelter, 1977, p.371).

As more people migrated to Toronto, rents rose and there was much subdividing of property. Thus, the boom also brought squalid conditions for many low income residents. Dr. Charles Hastings, appointed chief medical officer in 1910, advocated a far reaching programme of action: improved water and sewage systems, social housing, hospitals, parks and playgrounds. In the years leading up to the war, mayor Horatio Hocken implemented many of Hastings' recommendations by raising local taxes and borrowing. The city centre also changed dramatically, with new office blocks, banks, hotels, and department stores. With few controls over traffic circulation or land-use, downtown was becoming congested. In response, the Civic Improvement Committee drew up an advisory plan which featured a sweeping 'King Edward' boulevard and a grand 'Federal Avenue'. The latter linked a new Union Station to a rebuilt City Hall flanked by gardens and a parade ground. Nothing was built, but the plan conveyed an image of the future which appealed to the city's paternalistic elite.

In this context, public transport was seen as an amenity which had prime importance to citizens and businesses. The City of Toronto wished to see it expanded and improved. The year 1891 saw the end of a franchize to a tramway company, founded by the pioneering British transport entrepreneur Alexander Easton. After an unseemly row over the price, the municipality bought the assets for $1.4 million, yet it remained in public ownership for only a few months. Lacking the capital to upgrade the system themselves, the council reluctantly granted another 30 year franchize. Under its president William Mackenzie, the Toronto Railway Company (TRC) electrified the network of streetcar routes, and at first demonstrated innovation and flair, designing and building its own fleet. Under the terms of the franchize, it ran a special car to water the streets and keep the dust down in summer. The TRC even trialed cars to deliver mail, collect rubbish, and carry freight. It also offered value for money, maintaining a five cent cash fare: "the Lowest Streetcar Fare on Earth!"

Despite the promise of the first decade or so, the TRC was showing signs of wear and tear by 1914. World War One had brought problems with shortages of materials and labour, as well as two fires which destroyed a third of the company's streetcar fleet. The TRC struggled to cope with an increasing ridership and overcrowding on its vehicles. Yet, the underlying issue was the company's reluctance to reinvest. Mackenzie had risked and lost much of his fortune on an unsuccessful transcontinental railway venture. Meanwhile, the streetcars became unreliable and overcrowded in the peak. The TRC ignored directives to run more cars under the terms of the franchize, opting to pay fines instead. Furthermore, the TRC refused to extend its lines beyond the 1891 boundary, despite the substantial expansion of the city and its population. The

City Council brought a case against the TRC for not extending services into the new and annexed areas. The court found in the company's favour, but the tension continued.

Mayor Hocken's successor Tommy Church (1915-24) brought a new emphasis on fiscal prudence which slowed down the welfare programme, scaling down the funds for education and social services. Nevertheless, as Lemon has shown (1985, p.43) mayor Church was quick to appreciate the electoral attraction of municipalization to improve transit services, especially for disaffected voters in badly-served areas. As the franchize for one TRC line expired, Church marched amid much publicity with a gang of council workers to tear up the tracks commenting that: 'The time had come to reassure restless residents in the north end, already tempted to secede, that they were part of the city and deserved better and cheaper transportation services'. Furthermore, the influential 'Toronto Star' newspaper lent its support to the cause of municipalization, arguing that service and dividends were synonymous since 'the people themselves are the shareholders'.

Public disenchantment with the unstable 'boom or bust' character of the private enterprize era mounted. And, as the franchize period drew to a close, the City Council held a referendum on the question of public versus private ownership. In 1920, Toronto voted overwhelmingly in favour of municipalization of all local street railways. Thus, with public support, the City Council lobbied the provincial government to legislate for the creation of the Toronto Transportation Commission (TTPC) in 1921. As with other commissions constituted to carry out special functions, the TTPC was governed by a board of three 'citizen commissioners', appointed by the City Council. Within the first few years, the TTPC embarked on a massive $50 million programme, substantially backed by City Hall. As Lemon comments (1985, p.43) the city's high debts during the 1920s were largely the result of their investment in the TTPC and in the municipalized hydro electricity company.

David Harvey, general manager of the TTPC from 1924 was a civil engineer whose serious, methodical and businesslike character personified the new ethos. He adopted the maxim 'it pays to modernize'. Although he shunned personal publicity, Harvey emphasized the value of good design and good public image. The TTPC acquired 575 Peter Witt streetcars, designed to offer extra comfort, speed and capacity, with all-steel construction and power-sliding centre doors to reduce dwell times. The fleet was painted in a bright red livery because 'no longer would streetcars slink along apologetically or unnoticed'. The inadequate maintenance facilities were replaced by a new depot on the site of the old racetrack on Bathurst Street. The TTPC rebuilt most of the TRC lines and added over 40 miles of new route. It united nine separate street railway networks with through ticketing, interchange having been problematic on some corridors under the old regime. Raising the single zone fare to seven cents, the massive loans were eventually paid off by the mid 1940s.

By the late 1920s, much hardship remained for working people and their families, but Toronto's reform programmes and commitment to public infrastructure distinguished it from other cities, particularly those on the American side of the Great Lakes. Commissioner of Finance George Ross reported in 1927:

> The City is prosperous and there is a strong community feeling, which has manifest itself in the municipal ownership of transportation, light, and power systems, waterworks and other public services. The citizens take pride in the splendid condition of their city, its clean, well lit streets and boulevards and its fine parks and recreation centres (Lemon, 1985, p.19).

Two years later, the Wall Street Crash led to widespread unemployment, and the TTPC's ridership dropped by over 20% from 194 million in 1928 to 148 million in 1933. Nevertheless, under David Harvey's careful management, the system survived in good shape, introducing measures such as one person operation of vehicles, and the replacement of a loss making line with buses.

As the North American economy recovered in the mid 1930s, the TTPC focused its marketing efforts on competition with the rising number of private cars in the city. Thus, it developed an innovative strategy which included the introduction of rear engined buses, and in 1938 acquired the first of its Presidents' Conference Committee (PCC) streetcars, with their advanced technology. World War Two brought a drop in automobile production, and shortages of petrol and tyres. This time, the public transport system was well placed to take advantage of the boom in ridership, which doubled from 154 million in 1938 to over 300 million. In the years after the war, as servicemen returned and a new wave of immigrants arrived, the ridership climbed to a record 314 million in 1949. In these favourable conditions, the TTPC generated a surplus every year from 1940 until the end of the decade, although the postwar revival of automobile production and use brought problems for TTPC streetcars on Yonge Street and other prime routes.

Toronto was poised for growth, at a pace not seen since the first decade of the twentieth century. During the war, the City Council had set up an advisory Planning Board to consider transportation and other public services along with housing, business area development and other strategic issues for the wider city region or 'Metropolitan Area'. In 1944, a 30 year Master Plan had been drawn up to accommodate the forecast increase in Toronto's population through suburban expansion, and the associated increase in car ownership and use. Unlike its prewar antecedents, it was more concerned with economic progress and demography than with civic grandeur as such. Nevertheless, it was lofty in concept and envisaged centralized planning on a formidable scale. The Plan played down municipal boundaries and its implementation would require a 'partnership' between municipalities.

Although, in practice, the Plan was toned down and scaled down, the new approach led to a two-tier system of local government. In 1953, Ontario enacted legislation to create Metropolitan Toronto (Metro) as an elected authority, responsible for planning, major roads, public transport, and other strategic functions, funded through a levy. The City of Toronto and 12 other municipalities (rationalized to six in 1967) retained responsibility for local matters such as residential streets, street lighting, and parking. The renamed Toronto Transit Commission (TTC) became answerable to Metro, whilst the number of commissioners was increased from three to five to include the Metro's Chairman, and one other council member (the citizen commissioners were replaced by three Metro councillors in 1988). The revised structure was designed to facilitate more comprehensive planning and co-ordination of land-use change and transport investment.

Transport for a modern city and its suburbs

The year 1954 saw the opening of Canada's first underground railway - a short route from Toronto's Union Station 4.6 miles north along Yonge Street to Eglington Avenue. A key element of the TTC and Metro's strategy, the subway was also a symbol of the city's economic and technological progress. Although the project had been discussed as early as 1910, the time was now right as employment in the central business district boomed, and traffic congestion worsened on the main thoroughfares, affecting the reliability and cost of streetcar operation. The TTC wisely decided to use the healthy surplus it had accumulated in the 1940s to finance construction of Yonge subway at a cost of $67 million. As intended, the subway, and its subsequent extensions, proved a highly effective catalyst for commercial development, bringing new employment and tax revenue to the city. The first section of an east-west line opened in 1966, also stimulating economic growth within walking distance of the subway stations (Shaw, 1993, pp.255-56).

The 1950s, nevertheless, brought considerable challenges for public transport in Toronto. With responsibility to serve the whole of Metro's population, the area covered by the reconstituted TTC increased from 35 to 240 square miles, and buses were needed, especially on lower density suburban routes. Between 1953 and the end of the decade, bus operations quadrupled to 30% of the route mileage. The population of the Metropolitan Toronto area increased rapidly from 1.2 million in 1951 to 1.6 million in 1961, a rise of 33%. The number of automobiles per 1,000 people in Ontario more than doubled from 80 at the end of the war to 180 by the late 1950s. Thus, rising affluence inevitably brought falling ridership and financial problems for public transport, as in every other North American city, '...by 1958, the lifestyle of Toronto residents was drastically changed compared to the 1940s. The automobile was an accepted part of middle class life, incomes and leisure time options had greatly increased, and transit was becoming the second choice for transportation, especially for off peak travel' (TTC, 1989a, p.6).

Furthermore, the city expanded well beyond the area of Metro's jurisdiction. With the city centre and inner suburbs already built up and densely populated, the surrounding region was developed to provide new, more spacious accommodation for car owning households. The case for urban expressways was strongly supported by business interests which dominated provincial and municipal governments. Metro, with its responsibility for arterial roads, formulated proposals for a system of urban expressways, derived from the 1944 Master Plan. In the 1950s and 1960s, the municipality successfully bid for a capital grant to construct radial highways east-west along the lakeshore, and north-south along the Don Valley, linking up with the federal highway. An 'inner box' of elevated highways was needed to complete the programme, and in 1970 Metro, with support from Ontario, had every intention of implementing the scheme.

Meanwhile, the TTC's period of self sufficiency was at an end. Although the first section of subway had been internally funded, subsequent extensions needed low interest loans or capital grants. Despite his wholehearted support for the urban expressways programme, Metro's first chairman Fred Gardiner recognized that their peak capacity would soon be filled. Thus, Gardiner lobbied Ontario for finance to enable TTC schemes to go ahead, although it was not until the 1960s that the province answered Metro's appeal. A formula emerged, which eventually resulted in a 75% capital grant from Ontario. The subway system, fully integrated with surface transit through well designed interchanges, and through ticketing, did much to arrest the decline of public transport. Nevertheless, after a period of financial instability in the 1970s, the TTC is now expected to recover 68% of its operating costs from the farebox, the subsidy being split on an equal basis between Ontario and Metro.

The TTC has therefore become increasingly dependent on public finance, in return for which it must fulfil a range of economic, social and environmental goals to support the policies and programmes of the municipality and the province. Given the political and administrative framework within which the TTC was constrained to work, however, considerable tension emerged between the interests of the city centre/prewar suburbs and those of the new suburbs built on greenfield sites. Sewell (1991, pp.31-38) provides some valuable insights into the polarization between:

* the 'Old City' of Toronto's downtown and inner neighbourhoods and industrial areas, mostly developed to be served by public transport. This core area of the metropolis has a characteristic grid-iron street pattern, with housing densities of around 20 units per acre and mixed land-uses; and

* the 'New City' of the surrounding postwar suburbs and regional subcentres, mostly built to be served by the private car. Suburbia and an exurbia of summer holiday homes and leisure facilities stretches out along the lakeshore and northwards, over 50 miles from the centre. Housing developments with densities of around eight units per acre, are laid out along discontinuous residential roads without sidewalks, segregated from other land-uses.

As in the United States, the differences in outlook between the Old and the New City became accentuated in the 1950s and 1960s. They were to find expression in the debates of Metropolitan Toronto, with some marked differences between policies advocated by suburban politicians and those representing constituents of the older centre. Unlike their counterparts in many American cities, however, Toronto's Old City includes some very fashionable and stable neighbourhoods. Fortunately, these did not become run down ghetto areas, nor 'twilight zones'. Gentrification has been an important factor in the social geography of Toronto and other Canadian cities, notably Vancouver and Winnipeg, following a pattern similar to large cities in Britain. From the mid 1960s, there was a counter-movement of people who rejected the whole idea of suburban living - bohemians, artists, and academics were followed by young upwardly mobile professionals. Termed 'urban adventurers' and 'white painters' in the local press, the gentrifiers chose to live in refurbished Victorian or Edwardian property, within or adjacent to working class neighbourhoods and older industrial areas. As Reid has emphasized (1991, pp.10-30), the presence of a new class of educated, articulate people with opinions and lifestyles very different to those prevailing in the postwar suburbs became a significant catalyst for reform and 'citizen' movements in Canadian urban politics in the late 1960s and early 1970s.

By the late 1960s, transport policy had become the subject of a major debate which divided the Old and the New City. Although the previous schemes had met with little resistance, three more 'superhighways' penetrating into the heart of the city provoked furious opposition. These included the Spadina Expressway - a proposal which would necessitate the demolition of over 1,000 homes. Residents of the working class and gentrified neighbourhoods which lay in its path, as well as local groups concerned with conservation of Toronto's architectural heritage, formed a broad coalition 'Stop Spadina -

Save Our City'. A persuasive critique of the motorway extension was published by Nowlan and Nowlan (1970) *'The Bad Trip: The Untold Story of the Spadina Expressway'*. Nevertheless, as Sewell observes (1993, p.178), those who lived in the outer suburbs had difficulty comprehending why anyone should want to save older neighbourhoods, and thought it entirely reasonable that the existing city be demolished to make way for the new. The suburban councillors and businesspeople who then formed the majority on Metro Council had given their full support to Spadina. Indeed, they considered an expressway system to be the quintessential feature of a modern city.

Four years of almost non-stop lobbying and demonstrations, however, resulted in the cancellation of Spadina Expressway by a new provincial government. In 1971, incoming Ontario Premier William Davis announced 'If we are building a transportation system to serve the automobile, the Spadina Expressway would be a good place to start. But if we are building a transportation system to serve people, the Spadina Expressway is a good place to stop'. The vitality of the city region could not, however, be jeopardized. Ontario therefore focused its attention on public transport to facilitate higher levels of commuting to the city centre. 'GO Transit' originated from a provincial transportation study in 1965 which recommended public finance for a commuter rail service to relieve congestion and 'reduce the need for costly construction or expansion of expressways' (GO Transit, 1992, p.6). Starting as a small scale experiment in the late 1960s, it has received 100% government funding for rolling stock, upgraded track, and suburban stations with extensive park and ride. Today the network comprises seven rail lines with feeder bus services, radiating 35 to 50 miles from Union Station.

The 'U turn' on Spadina Expressway also helped to shift the balance of power at Metro and reinforce support for public transport in Toronto itself. A group of left-of-centre councillors associated with the anti-motorway campaign had also helped to expose the complicity of Metro's then leaders in some highly unpopular redevelopment schemes. A large turnout of voters in 1972 elected a reformist council led by mayor Crombie. Although the radical policies lost momentum later in the decade, investment in the TTC received a major boost which continued through to the scaling down of public expenditure in the early 1990s. In 1972 citizens protests had reversed a previous decision to abandon the streetcar network (Wickson, 1991, p.63), and the TTC began to upgrade and extend the seven remaining lines, with new, European-style streetcars introduced from 1978. In accordance with Metro's long range plan to attract inward investment to the central business district and new strategic subcentres, the subway lines were extended north, north west and west with extensive parking at key stations. In 1985, a state-of-the-art automated light rail line was opened to Scarborough, 12 miles to the east. And, in 1990, a light rapid transit line was opened to connect downtown with the Harbourfront regeneration scheme in the former docklands area.

Thus, considerable public resources have been allocated to establish Toronto as a 'world city'. Since no attempt has been made to revive the urban motorway programme, public transport has benefited with 75% grants from Ontario for extensions to the subway and light rail schemes, as well as 100% grants for new GO Transit commuter lines serving the wider city region. Such prestigious schemes have been carried out alongside other 'grand projects', including the SkyDome stadium, and infrastructure to facilitate Metro's three large scale commercial developments: the 'new downtown centres' in Scarborough and North York, and the redevelopment of Harbourfront. In the context of competition with other North American cities in an increasingly globalized economy, the strategy has been highly successful in attracting new 'idea driven' industries and creating a civilized 'liveable environment' for business and leisure (Shaw, 1993, p.256). Thus, for almost 25 years there was a remarkable consensus across the mainstream political establishment of the municipality and province, supporting an ambitious programme of pump-priming investment. Although rising public debt forced a slow down in the early 1990s, the neo-Keynsian strategy was halted only in 1995 with the election of a far-right Conservative Government in Ontario.

Transport, popular dissent and demands for social equity

The prevailing drive for economic growth, and a city of world status has not, however, been without its critics. The popular dissent of the early 1970s and the election of a reform-minded mayor had shown that the established civic and business leadership could be challenged on critical issues such as the urban motorway programme. Over the past 20 years or so, 'community politics' has also been expressed in well articulated demands for social equity and equal opportunity, especially regarding the provision of Metro services. With regard to the TTC and GO Transit, the debate has not been about the need for investment in public transport as such. Rather, it has focused on the priorities favoured by the provincial and local government, and their failure to appreciate the social dimension of their decisions. Some community groups have argued that investment in prestigious subway, light rail and commuter lines bring greatest benefit to real estate developers, city centre employers, and to more affluent long distance commuters. Consideration should have been given to alternative policies with an emphasis on creating a safer city, accessible to all Torontonians, regardless of disability, gender, race and so on. Programmes to upgrade local bus and streetcar services, for example, could bring greater benefit to those most dependent on public transit. By the 1990s, this had led to a fundamental review of the agenda for public transport and other services, a debate which still continues in the context of reduced resources for all public programmes. Demands for service equity have been expressed by three major 'identity groups' - disabled people, women, and ethno-racial minorities.

Service equity and disabled people

Organizations representing disabled people in Canada have been a potent force for change. In the 1970s, they gained encouragement from the successful outcome of campaigns led by Vietnam War veterans in the USA. Yet, whereas their counterparts in the United States argued that all buses should have lifts to allow access for people in wheelchairs, in Canada they set a different agenda. Instead, they lobbied for 'parallel' transit, to provide door to door access in special vehicles. In Toronto, disabled rights activist Beryl Potter, first set up a demonstration project using a small fleet of converted vans driven by volunteers in her neighbourhood. Mrs. Potter then led a campaign to persuade the TTC to take over responsibility for parallel transit throughout Metro, with purpose built vehicles driven by professional drivers paid the appropriate union rate for the work. As a result of effective lobbying by voluntary organizations representing Toronto's disabled communities, positive media coverage of the issues, and support from the reformist Metro Council and the Amalgamated Transit Union, the campaign achieved its goal. In 1975 the TTC began its 'WheelTrans' operation using nine vehicles, and carrying 22,000 people in its first year. Two decades later, WheelTrans has a fleet of over 130 purpose-built Orion II air-conditioned vehicles carrying around 1,320,000 people per annum. There is also an extensive taxi service for less severely disabled people, and a fixed-route wheelchair accessible community bus network.

During the 1980s, however, attention shifted to mainstream public transport in Canada, as disability groups lobbied federal and provincial governments to make conventional bus, streetcar and rail services more accessible. There was much criticism of service providers for past mistakes and omissions - their failure to design public transport which could be used by everyone. In response to high profile campaigning, Ontario province introduced its 'Easier Access' programme in 1990. Implemented by the left-of-centre New Democrat Government, it provided enhanced funding for a range of mobility improvements to the TTC subway and to GO Transit commuter rail. Thus, from 1990-95, the TTC received a 90% grant from the provincial Ministry of Transportation to retrospectively fit 23 key interchange stations with easier access features. The improvements included lifts to provide an alternative to escalators, ramps, handrails, and tactile flooring to assist blind and partially sighted passengers. With regard to surface transport, recent developments in vehicle technology offer significant opportunities. The new generation 'low floor' buses and streetcars can 'kneel' to a level which is accessible to 90% of ambulant disabled people, and are fitted with user-activated wheelchair ramps. Since entry and exit is easier and quicker for all passengers, including parents with children and push-chairs, dwell times at transit stops are reduced to everyone's benefit. In 1993, Ontario became the first province to require municipalities (including the TTC) to purchase low floor vehicles, by making them a condition for receipt of the 75% capital grant.

As disabled people's demand for personal mobility increases, the argument for making conventional transit fully accessible become more compelling for service providers such as the TTC. From a financial point of view, the very popularity of the parallel WheelTrans service now poses serious problems for the TTC. Each journey costs more than $40 (£20), yet the fare to seniors is less than $1. Bob Evans, the unit's director, commented (personal interview, 1993) 'It is very evident that the WheelTrans service as it exists today is unaffordable. We are going into a situation over the next three years when very little money will be made available to us, and yet we have a ridership increase of between 8% and 12% annually'. The current action plans which are reducing the physical barriers to mainstream transit should help relieve the pressure on WheelTrans. '... So, the strategy is to make the conventional system fully accessible, with the expectation that people who presently have difficulty climbing steps will use the elevators to get into the system, and some in wheelchairs too'. Even with the 'carrot' of enhanced funding for the 'Easier Access' programme, and the 'stick' of making low floor vehicles mandatory, it will, however, take many years to convert the entire TTC system. The consensus among disability groups in Toronto has, nevertheless, swung in favour of making the TTC fully accessible over the coming years.

First, the principle of integration is preferred to the segregation of disabled people in the provision of public transport as in other services such as education and leisure facilities. Furthermore, despite investment in sophisticated computer technology in the WheelTrans booking system, the service cannot offer the same spontaneity and range of travel opportunities as the conventional transit network. Thus, disability groups support the TTC's policy of creating a 'family' of transit services in order to create a range of options which take account of the full spectrum of mobility needs. The need to ensure "dignity and respect" for the user, and the full participation of user groups in policy and practice must be guiding principles. As with the approach to personal security discussed below, a 'partnership' approach has been favoured. As an outcome of a collaborative working party with representatives of the local disability group Trans-Action, the Government of Ontario, Metropolitan Toronto, the TTC, and the Amalgamated Transit Union, a set of recommendations - 'Action for Access' was drawn up. These included a permanent forum to work out joint solutions to problems in the critical years ahead. Thus, in 1993, the TTC established the 'Advisory Committee on Accessible Transportation' with 15 members to articulate the views of the various disability 'constituencies' including the blind and partially sighted, the deaf and hard of hearing, people with multiple sclerosis, and so on. Its members are appointed as individuals, rather than to represent organizations, and each serves for a three year term of office.

Service equity and women

Although reported incidents of violent crime are much lower than in the United States, many women in Canada have justifiable fears of being assaulted or harassed while using urban public transport. Such anxieties, felt by other groups such as the elderly and ethno-racial minorities, may deter some from travelling at all, especially after dark. In Toronto, concern for the personal safety of women and children was heightened in the early 1980s. An horrific series of rapes and murders in the city and the surrounding suburbs led to widespread public outrage. The 'Women's Movement' in Canada was already well organized, and a number of local groups came together, demanding action from Metro and its Commissions, including the Police and the TTC. The latter already had a long standing commitment to high standards of safety and security, but as several women's groups emphasized, previous safety audits had focused on violent robberies, and had not specifically addressed the issue of sexual attacks. Following the recommendations of a special working party, the Metro Action Committee on Public Violence against Women and Children (METRAC) was set up in 1984, as a voluntary organization, part funded by but independent from Metro.

Working with Professor Wekerle and her colleagues at the nearby York University, METRAC developed a set of original methodologies for assessing the personal safety hazards of public places, as well as generating ideas for improvement. Then in 1987, following an approach to Alan Leach, the newly appointed chief general manager of the TTC, the techniques were tested and developed in the context of Toronto's public transport system. Thus, a unique collaborative project was initiated by METRAC, the Police and the TTC to improve personal safety. Despite their common goal, it was recognized at the outset that the three organizations did not share any common philosophy or ways of getting things done. Traditional police crime prevention programmes had not included sexual assault prevention; transport undertakings had not focused on sexual assault; women's groups tended to distrust traditional male corporate culture, and had rarely been on close terms with police or transit organizations. Working together would not be easy. Nevertheless, as the project developed, '..all involved found that our effort to make transit safer for women was better because it was a joint venture' (TTC, 1989b, p.6).

An important feature of the collaborative project was the audit process, which was informed by surveys, public meetings and focus group discussions. Feedback from women passengers was sought on a step-by-step basis. As a result, a number of modifications were made before the final action plan was approved in 1989. The centre piece of the programme of environmental improvements was the installation of 'Designated Waiting Areas' (DWAs) in all subway and light rail stations by 1993. These provide a safe, well lit location on every platform, allowing passengers two way communication with nearby staff, who are trained to provide assistance. Members of METRAC were fully involved in piloting the scheme. Three prototype DWAs were tried out, and METRAC's preferred option was chosen, even though it cost more than the other designs. Another important change is that the plans for all new and refurbished subway stations are now scrutinized by an architect whose responsibility is to examine proposals from a personal security perspective, in consultation with any local women's groups. The current postholder is a woman whose work has been praised by METRAC. Smaller scale physical improvements, which have been carried out systematically following the audit process, include better sightlines, mirrors on corners, clearer signage of exit routes, and features such as extra benches and public art to 'humanize' the subway and bus/streetcar interchanges. Many of the women involved in the security audit process, nevertheless, emphasized the need to go beyond purely physical measures to 'design out crime'.

Complementary aspects of the action plan drawn up through the 1986-89 collaborative project have therefore included staffing measures, especially training customer contact staff. In focus group discussion, and in surveys, women had stressed that TTC employees needed instruction on how to prevent or respond to incidents of sexual assault. These must include harassment, such as unwanted touching, being followed or verbally abused. It was recognized that the staff could be placed in potentially dangerous situations, and might be uncertain of their role. It was also noted that the TTC's equal opportunity programme was successfully recruiting more female employees for these and other positions. Since the early 1990s, all staff have received instruction on how to respond quickly and sensitively, including advising victims where to seek help. METRAC and other groups also advocated the need for public education to inform passengers about the DWAs and other security features. In response, the TTC's equal opportunity and marketing departments have jointly produced and distributed leaflets about these issues in 15 languages, and since 1995 provide multi-lingual disks for use by community organizations. A further example is the wording of the decals for passenger alarms, which now include 'harassment' as specific reason for communicating with the guard. Initiatives on surface transport include the 'Request Stop Programme' which enables women and others travelling alone on TTC vehicles after dark to alight at places other than official stops.

Service equity and ethno-racial minorities

As Qadeer (1994, p.188) comments 'multi-culturalism is a fact of life in Canada, though this has been recognized only in recent times'. The federal government's Immigration Act, 1967, was an important watershed, since it ended national quotas and introduced new criteria covering education, skills, family relations, and more latterly refugee status. During the 1980s, about two thirds of new Canadians came from Asia, Africa, and Latin America, Toronto being a prime destination, along with Montreal and Vancouver. Metro Toronto has a rich diversity of cultures and languages with over 60 communities or 'neighbourhoods'. More recent arrivals include Cambodians, people from the Caribbean, Ismaili Muslims, Pakistanis, Sikhs, Sri Lankans, and Vietnamese. Census data shows that between 1986 and 1991 the proportion of Metro Toronto's population with neither English nor French origins rose from 45% to 59%. Ethno-racial minorities, especially new Canadians, are less likely to have access to a car, and therefore tend to be particularly reliant on public transport. Language differences have increased significantly. Residents of Toronto who use a mother tongue other than English at home, now number almost half a million (about a fifth of the population) - a significant market for the TTC. Some neighbourhoods are very important generators of revenue on the transit lines which serve them. For example, route 77 through Chinatown, with its 200,000 residents is the TTC's most profitable service.

External pressure to develop ethno-racial access policies and initiatives came initially from Metro which has the power to issue 'directives' to the TTC, and from the 'Community Reference Group' (CRG). The latter is funded by the municipality and functions as an advisory panel with its positions advertized and filled according to qualifying criteria. As such, it is not regarded as the definitive voice of the city's communities, but plays an important role in informing the policies of Metro and its various commissions such as the TTC. It also provides linkages and useful channels for communication. For example, a member of the CRG approached the TTC to ask for increased capacity when the Muslim community celebrates the Eed-Ul-Adha. The TTC now runs special services to transport the city's 25,000 Muslim worshippers to mosques and other places of prayer. Furthermore, the electronic destination screens on the vehicles are used to inform the wider public of what is happening, as they do with secular events such as baseball matches. Similar liaison between the TTC and community organizations now takes place over other, large scale events, notably the Chinese Lantern and Dragon Boat Festival and Caribana, both of which have become major tourist attractions, attended by over a million visitors. As Ron Chong of the equal opportunity department commented (personal interview, 1993) 'There is not only a moral dimension but also a revenue issue, making people feel more welcome'.

Although the issue of 'ethno-racial access' did not feature on the agenda of transit undertakings in Canada until the mid 1980s, it is now accepted as an important aspect of corporate strategy for operators in cosmopolitan cities such as Toronto. In 1992, the TTC adopted a policy which commits the organization 'to identify and eliminate barriers in order to ensure that diverse ethno-racial and linguistic communities have access to information and transit services, and to the planning of services provided by the TTC' (TTC, 1992, p.10). An action plan to implement the policy was developed by the TTC's equal opportunity department, which recognized that this would require a comprehensive approach. Initiatives currently being pursued include: multilingual interpretation and translation services; outreach and information; staff training and development; collection of ethnic minority and Aboriginal data; increased community linkages; programme review and development and monitoring. For example, the TTC's most widely read information brochure - 'Your TTC' - provides basic information for people unfamiliar with the system, including an explanation of how to use safety and security features such as the DWAs described above. The brochures are now available in 15 languages, along with 'Project Access' software information, through local community organizations.

Metro and the CRG recognize that the TTC, through its equal opportunity department, has taken a proactive approach. Nevertheless, there is little room for complacency. As a TTC manager commented (personal interview, 1995) 'it is no secret that we did have some human rights related customer complaints and they were on the increase'. There were also allegations of racially insensitive behaviour by TTC employees, which were referred to the CRG, and hence entered the public domain. In the early 1990s complaints were documented concerning alleged harassment of East Asians, particularly the elderly, by streetcar drivers on routes through Chinatown. It was alleged that drivers were over officious and sometimes abusive. The matter was taken up by a long established Chinese community organization, who discussed the grievances at a specially convened meeting of the Commission. This led to the introduction of new complaints procedures, including an interpretation and translation service, allowing non-English speakers to lodge complaints and suggestions in their mother tongue. The TTC have also recognized that customer contact staff should also receive training. In 1995 they launched 'All Aboard: Respecting Diversity at the TTC'. This new one day programme focuses on service equity and on increasing awareness of race relations issues. The equal opportunity department hope that finance will allow it to continue, reaching all 4,500 customer contact staff in four years.

Metro's Official Plan (Metro, 1994, p.41) emphasizes the municipality's policies 'to achieve strong communities where diversity is valued and residents have equitable access to services and opportunities, where the public realm is usable, safe, recognizable and appealing, and where residents are able to participate in decisions affecting them'. Thus, Metro Toronto services will focus their efforts on: 'removing ethno-racial, linguistic, physical or other systemic barriers to services and resources'. It is only in recent times, however, that such issues have been given priority. Initially, the voluntary organizations who campaigned for a more accessible and safer city received little or no coverage in the news media, and found it difficult to communicate their proposals to policy makers and the wider public. After many years of lobbying and discussion, they have gained recognition and support from Metro, as well as from sympathetic individuals within organizations such as the TTC. Today, formal procedures for consultation are in place, conferring credibility and respect for the views of passengers who are disabled, female, and/or members of ethno-racial minorities.

Conclusion

Unlike its counterparts in Britain, Toronto's public transport system remains a fully integrated municipal service with the privilege of monopoly over the geographical area it serves. Following an unsatisfactory experience of franchizing to private street railway companies, the transit system was acquired by the city government in 1921. Since that time, local elections have brought some significant changes of policy and approach. Nevertheless, the essential principle of municipal ownership and accountability has been supported for three quarters of a century. The Toronto Transit Commission's motto: 'Service, Courtesy, Safety', was adopted to express the mission of a high profile public utility with a key role to play in the life of the city. Reference is still made to these core values in the undertaking's publicity material, as well as its internal marketing, including the training and socialization of new employees.

Today, however, municipal undertakings must apply such principles to the requirements of a sophisticated urban population which is socially and culturally diverse. Consumers of public services in Canada are increasingly assertive and well able to organize themselves into powerful lobby groups (Shaw, 1995, pp.40-59). They expect value for money and clear benefits, not only for fares paid, but also for subsidies funded through local taxation. Thus, the transit undertaking's long range policy statement for the 1990s reaffirms its core values adopted 75 years ago, yet it also admits that: 'Over the past several years, the TTC has witnessed a number of social, economic and political changes that will affect its role and mandate in the community ... pressure for the TTC to assume a more social role in providing transit service, ... a perceived lack of openness to the community it serves' (TTC, 1989c, p.1).

For a municipality positioning itself as a 'Liveable City' for the twenty first century, a high quality public transport system is a key element of the strategic vision: 'Metro residents use the parks, roadways, and transit system daily ... They contribute to a sense of civic pride and community identity' (Metro, 1992, p.73). In some respects, the rhetoric is reminiscent of that used by the City Beautiful movement, with its image of orderliness and beauty in the provision of public space and facilities. Unlike the vision of a century ago, however, there is an emphasis on valuing a diversity of lifestyles and cultures, rather than conformity imposed from above. In Canada, human rights legislation and policies to promote equal opportunity help to secure commitments from public services to become more accessible, and responsive to the disparate needs of society. At local level, client groups now negotiate with service providers to ensure that social equity is a key feature of programmes and action plans. Furthermore, they are participating in a highly practical way in the implementation of these policy commitments to improve the quality of urban life.

References

Adams, T. (1915), 'Experience Under the English Town Planning Act' in *Proceedings of the Sixth Annual Planning Conference'*, Toronto, quoted in Gunton, T. (1991), cited below.

Artibise A. F. and Stelter G. A., (eds.) (1977), *'The Canadian City'*, McLellan & Stewart, Toronto, in Sewell, J. (1993), cited below.

Artibise A. F. and Stelter G. A., (1981), 'Conservation planning and urban planning: the Canadian Commission of Conservation in historical perspective' in Kain, R. (ed.) *Planning for Conservation*, Mansell, London.

Brownell, B. (1980), 'Urban planning, the planning profession and the motor vehicle in early twentieth century America' in Cherry, G. (ed.) *Shaping an Urban World*, Mansell, London.

GO Transit, (1992), *Twenty Five Years on the Go*, GO Transit, Ontario.

Gunton, T. (1991), 'Origins of Canadian Urban Planning' in Gerecke, K. (ed.) *The Canadian City*, Black Rose Books, Montreal.

Jacobs, J. (1965), *The Death and Life of Great American Cities*, Penguin, Harmondsworth.

Kiepper, A. F. (1994), *Meeting future mobility needs with yesterday's transit infrastructure*, London Transport, 10th Annual Lecture, London.

Lemon, J. (1985), *The History of Canadian Cities: Toronto Since 1918*,

Lorimer & National Museums of Canada, Toronto.

Marcuse, P. (1980), 'Housing policy and city planning: the puzzling split in the United States 1893-1931' in Cherry, G. (ed.) op. cit.

Marsh, B. C. (1908), 'City Planning in Justice to the Working Population, Charities and the Commons' quoted in Wilson, W. H. (1989), cited below.

Metropolitan Toronto, (1992), *The Liveable Metropolis: Metropolitan Toronto Official Plan* (Draft), Toronto.

Metropolitan Toronto, (1994), *The Official Plan of the Municipality of Toronto*, Toronto.

Nowlan, D. and N. (1970), *The Bad Trip: the Untold Story of the Spadina Expressway,* New Press/House of Anasi, Toronto.

Qadeer, M. (1994), 'Urban planning and Multiculturalism in Canada' in Thomas, H. and Krishnarayan, V. (eds.) *Race Equality and Planning*, Avebury, Aldershot.

Reid, B. (1991), 'The Story of a New Middle Class' in Gerecke, K. (ed.) *The Canadian City,* Black Rose Books, Montreal (originally published in City Magazine, Spring 1988).

Robinson, C. M. (1913), 'The Improvement of Towns and Cities or the Practical Basis of Civic Aesthetics' in Lubove, R. (1967), *The Urban Community: Housing and Planning in the Progressive Era,* Englewood Cliffs, New Jersey.

Sewell, J. (1991), 'Old and New City' in Gerenke, K. (ed.) op. cit. (originally published in City Magazine, Spring 1988).

Sewell, J. (1993), *The Shape of the City,* University of Toronto Press, Toronto.

Shaw, S. J. (1993), *Transport: Strategy and Policy,* Blackwell, Oxford.

Shaw, S. J. (1995), 'Transport and the Assertive Consumer' in McConville, J. and Sheldrake, J. (eds.) *Transport in Transition,* Avebury, Aldershot.

Smith, P. J. (1979), 'The principle of utility and the origins of planning legislation in Alberta' in Artibise, A. F. and Stelter, G. A. (eds.) *The Usable Urban Past: Politics and Planning in the Modern Canadian City,* MacMillan, Toronto.

Toronto Transit Commission, (1989a), *Back to Basics: A TTC Strategy for the 1990s,* Toronto.

Toronto Transit Commission, (1989b), *Moving Forward: Making Transit Safer for Women,* Toronto.

Toronto Transit Commission, (1989c), *Transit in Toronto: the story of public transportation in Metropolitan Toronto,* Toronto.

Toronto Transit Commission, (1992), *Ethno-Racial Access Plan,* Toronto.

Wickson, E. A. (1991), 'Toronto's New Light Rail Line' in *Light Rail Review 2,* Light Rail Transit Association, London.

Wilson, W. H. (1989), *The City Beautiful Movement*, The Johns Hopkins University Press, Baltimore.

5 Investing in transport - who pays?

Martin Higginson

Background

This chapter considers policies for investing in city transport in Britain over the last 100 years, with particular reference to public transport. The principal phases of new urban transport investment, which display considerable overlap, are summarized in Figure 1.1.

Period	Mode	Investor
1890s	railways	private
mid 1890s to 1920s	tramways	public/some private
1900s:	London underground	private, UK and USA
1920s to 1950s	buses	private/public
1950s to mid 1990s	cars and highways	private/public
1980s to 1990s	metro/light rail	mainly public

Figure 1.1   **Period, mode and investor of principal phases of urban transport development**

However, the summary requires qualification in several respects. In addition to initial investment in new routes and modes of transport, there has been substantial capital investment in modernizing and expanding infrastructure and vehicles, including:

* rail electrification, which has taken place throughout the period under review, albeit with extended gaps mainly associated with the two world wars. Electrification began in the 1900s with suburban routes and, since the 1920s, has increasingly been extended to longer distance commuter and trunk lines;

* expansion of the London underground, principally suburban extensions since the 1930s, with extended gaps between individually large projects, and the construction of new lines through central London since the 1960s;

* highway construction and mass car ownership with some new highway construction and the beginnings of mass car ownership evident between the wars.

Second, the dates refer only to the main periods of new investment and expansion. In each case investment, especially in infrastructure, is long lived and the period of use of the new infrastructure extends far beyond the period of initial expansion, in many cases up to the present time. Examples of shorter lived investment include:

* conventional tramways, which had all but disappeared by the early 1960s (Glasgow's last routes closed in 1962, leaving only the Blackpool coastal line);

* rail closures in metropolitan areas, which have principally involved duplicatory or minor lines.

Third, the categories of investment listed in the table are by no means equal in value or importance:

* by the 1890s most construction of new railway links had been completed. Much of the investment during this decade was in widenings and other major improvements;

71

* investment in underground railways took place mainly in London, with much smaller developments in Glasgow and Liverpool;

* construction of metro and light rail systems since the 1980s has taken place only on a limited scale, but is significant because it represents a fundamental new direction in recent British urban transport investment.

The addition of new transport infrastructure and passenger services over the past 100 years has often taken place without the abandonment of existing facilities. This has brought about an overlay of alternative modes, principally general purpose highways, railways and motorways. It has also enabled 'metropolitan' travel to extend far beyond the areas formerly covered. The effects have included: an expansion of the built-up area; the phenomenon of living outside the continuous built-up area and commuting into it; and, partly as a consequence of these developments, reduced inner city population densities and improved living conditions. For example, the Metropolitan Railway combined the creation of new travel opportunities with the development of desirable residential communities in the catchment area of its services in and beyond north west London. The company mounted a prolonged campaign to promote living in "Metroland". Electrification of suburban railways, especially in south London under the "Southern Electric" banner, similarly generated the development of new residential areas (Jackson, 1973, pp.273-90; 1978, pp.266-326). To the north of London, neither the Great Eastern nor the Great Northern Railways, nor the successor London and North Eastern Railway (LNER), could afford to electrify their suburban routes. Instead, intensive steam operated suburban services were operated, most notably the "Jazz" services between Liverpool Street and north east London. Electrification was only eventually carried out in this area with the aid of government financial guarantees and support, under the 1930s New Works Programme and subsequently.

Freight movement

The model of overlays of transport, however, only relates to passenger movement. The picture is different for freight movement, where the old pattern of centrally located railway goods depots, serviced by short distance road cartage, has largely been swept away. Goods delivery by road throughout has become the norm for most traffics. Typically road freight movement is now inwards from outside the urban centres. The loss of city centre freight termini, often close to the main passenger stations, makes it impossible effectively to contain the flow of heavy goods vehicles on radial routes to and from cities. Even if a political decision were made to try to effect a return to rail freight movement, its implementation would be hard to achieve as the urban and freight transport economy have changed so substantially. Many of the former depot sites have been redeveloped, traditional businesses have relocated and reoriented their transport requirements from rail to road, and the nature of freight movement to and from city centres has changed fundamentally.

Society at large has paid for freight users' pursuit of cost efficiency, with insufficient regard paid to externalities, and for the railways' inability to adapt their services to meet consumers' changing demands. Although short distance rail freight traffic, except in bulk, is generally uneconomic, the loss of many longer distance flows to road should have been avoidable. The North Eastern Railway, for example, was one of the most progressive companies in the early part of the twentieth century, investing in high capacity (40 ton) bogie freight wagons as early as 1903, long before the emergence of the large road goods vehicle, but their example was not followed (Tomlinson, 1914, pp.727-28). Even as late as the 1960s, the British Railways Board (BRB) was still building new four wheeled, short wheelbase freight wagons (British Transport Commission, 1961, p.31).

When the Freightliner concept of carrying international standard sized containers on dedicated trains was introduced in the mid 1960s, railway management failed to negotiate access arrangements satisfactory to the unions. Consequently, private road hauliers were initially not allowed into Freightliner terminals and it was not possible to provide a properly integrated road-rail service. BRB also took too long to arrive at an effective policy for providing less than train load freight services, with the result that by the time the Speedlink network of fast wagonload freight trains was launched in 1977, too much of the potential traffic had already been lost. Speedlink lost money heavily and despite the network's withdrawal in 1991 (BRB, 1992, p.20), it was found necessary to write off £500 million of the parent company Railfreight Distribution's debt in 1996.

Urban railways

By the end of the nineteenth century, the mainline railway companies had largely completed their networks from, to and within Britain's cities. The railways were private companies, although subject to government regulation, including controls over rates and charges (Savage, 1966, pp.70-82). An Act of Parliament was required to sanction the construction of new lines, but the government did not fund investment. The incentives to new construction provided by the Light Railways Act, 1896, did not extend to public funding. As Sherrington observed, the construction of small branches under that Act 'must have resulted in severe losses to the railway investor' (Sherrington, 1934, p.266), but this observation probably did not apply to the large number of urban tram lines constructed under the 1896 Act. Between 1897 and 1980, a third of the 300 Light Railway Orders passed under this Act were for urban tramways (Bosley, 1990, p.45, pp.182-91, App. A).

The last years of the nineteenth century and first of the twentieth were the heyday for local rail travel on the mainlines within Britain's cities. However, the arrival of effective tramway, motorbus and, basically in London, underground railway competition signalled the end of the railways' period of supremacy in this market, for which they were not, in any case well suited. Despite their coarse networks, mainline railway companies had attempted in the days before motorized road transport to provide internal networks within cities. The circuitous routing (involving, for example, services from north London to Victoria and from Croydon to Liverpool Street) and low frequency of many services demonstrates that they were not always well suited to this task. As soon as road and underground competition materialized, providing direct routes, high frequencies and stops closer to where people wanted to go, many such services withered away. The inner suburbs of most of Britain's cities contain examples of long-closed stations, such as Coborn Road (Great Eastern), and Haverstock Hill and Camden Road (Midland), in London, and Holbeck in Leeds. Complete suburban routes in the inner suburbs and between smaller centres gradually reduced in number. For example the Blackwall and North Greenwich branches, in east London, closed in 1926 and the line between Bradford, Halifax and Keighley via Queensbury in 1955. Heavily populated industrial areas such as the (former) Notts-Derbyshire and South Wales coalfields have seen once dense railway networks drastically reduced, where heavy goods traffic once helped to sustain local passenger services. Nevertheless, despite their wholesale withdrawal a generation and more ago, recent years have seen a minor revival of provincial urban and passenger railways. This has been influenced by two main factors; excessive road traffic and a growing interest by local authorities in railways now that their powers to plan bus services have been reduced. Nottingham's Robin Hood line, the Cardiff City line and the Swan line services around Swansea exemplify such reopenings in medium density urban areas.

Amalgamations and working agreements between companies also took their toll of some hitherto competitive rail services. The 1899 agreement between the South Eastern and London, Chatham and Dover Railways, thereafter run by the South Eastern and Chatham Managing Committee, for example rendered the Greenwich Park line redundant, as the direct Greenwich-London service was now provided from within the same organization (Connor, 1996, p.29). After World War Two, public ownership of public transport in Britain reached its maximum level and most of the system remained in the ownership of central and local government until moves towards privatization began in the 1980s.

Coverage of investment costs gradually became a clouded issue, as first the railways and then the buses slid into deficit. Initially capital debt write-offs and, subsequently explicit financial support, increased the public funding of public transport. The distinction between capital and operating support was not always as clear as it should ideally have been. The distinctions made in the early 1960s by Dr. (later Lord) Richard Beeching, chairman of the BRB, 1961-65, between profitable and loss-making railway routes and traffics mark a watershed in public transport costing and finance.

The Beeching era saw withdrawals of suburban services in many parts of the country. The catalyst for this development was the publication in 1963 of the famous 'Beeching Report' under the title *'Reshaping of British Railways'* (BRB, 1963), which identified the financial performance of each service and numbers of passengers carried. Although some of the costing may not have been accurate, for example with regard to the extent of contributory revenue from branches to mainlines, the underlying message of Beeching's value for money audit was incontrovertible. The report showed that large number of lines were carrying very small numbers of passengers and were failing to make a worthwhile contribution to British Rail's revenue. Although it may be argued that several 'mistakes' were made in deciding which lines to close (for example Oxford-Cambridge), the level of rail passenger usage only fell slightly. From 36 billion passenger kilometres in 1963, ridership fell to a low of 34 billion in 1968, which was followed by a recovery back to 36 billion by 1974.

A worthwhile and lasting outcome of 'Beeching' was the development of a system of subsidies for uneconomic lines whose social value justified their continued operation; this was implemented by the Transport Act, 1968, section 39. The busy North London line between (then) Broad Street (now North Woolwich) and Richmond was one such route, which was proposed for closure by Beeching, but subsequently reprieved. The policy of attempting to attribute financial outcomes to individual routes was replaced by a global Public Service Obligation grant under the Railways Act, 1974, which gave effect to European Community regulation 1191/69.

Only in the metropolitan areas where, under the Transport Act, 1968, (sections 10(1)(vi) and 20), Passenger Transport Executives (PTEs) were given the power to support local rail services, was local financial support data still publicly available. Between 1983-84 and 1993-94 total annual payments to the BRB by the PTEs ranged between £92m (1987-88) and £129m (1983-84), with an average of £108m, at constant 1993-94 prices. On average, PTE support amounted to 9% of British Rail's annual grant income over the same period (Department of Transport, 1995, p.30). Whilst PTE funding is usually limited to infrastructure investment, concessionary and operational support, limited purchase of rolling stock has also been undertaken, for example of class 158 diesel trains by West Yorkshire PTE. Some county councils have also invested in local railways, including service enhancement, reopening and station improvement projects.

More recently, disaggregated financial outcomes for individual parts of the rail network have again been made public, as train operation has been franchized as part of the process of rail privatization. More accurate revenue apportionment is now possible than in the 1960s, due to the wider availability of computer-based management information. Nevertheless, the allocation of joint costs and the apportionment of revenue between operators, services, times of day, days of the week, etc., still gives rise to controversy. The central issue is to decide which traffic should bear the full infrastructure cost and which may be carried at marginal cost. The committee of enquiry into the finances of the railways headed by Sir David Serpell reported in 1983. The report 'Railway Finances' (Department of Transport, 1983, para.13.18-13.21) discussed the principle of contribution accounting, whose adoption was an important feature of British Rail's reorganization into business sectors (passenger sectors were InterCity, Network South East and Regional Railways) in the 1980s.

Prior to the formation of Railtrack, three rolling stock leasing companies and the 25 train operating companies under the Railways Act, 1993, the infrastructure, stock provision and train operation had remained largely in the hands of single organizations. The main exceptions were in the conurbations where the PTEs assumed responsibility for local train service planning in their areas under the Transport Act, 1968, section 20.

Public finance for city centre rail distributors

Public finance has enabled many of Britain's principal cities to develop new or reopened city centre and cross-city rail links over the past 30 years. Liverpool, Glasgow and Newcastle each gained new or reopened cross-city railway distributors in the 1960s and 1970s. In Birmingham, the Cross-City Line, linking Redditch with Lichfield, opened in 1978. Several new stations were opened, but no new route constructed, to create this functional, but operationally difficult link. In London new construction has included the Victoria and Jubilee underground lines, the Docklands Light Railway and the mainline Thameslink, all publicly funded. In the 1990s, private finance has begun to make a contribution. State grant aid to the Manchester Metrolink was only 38% (plus 10% from European Development Fund), and for Sheffield Supertram 50%. The developer, Tramtrack Croydon, is making a 'significant financial contribution' to Croydon Tramlink (Hansard, 23 July 1996, cols.147-148). The privatized British Airports Authority (BAA) initially shared the cost of building Heathrow Express, the Paddington-Heathrow Airport railway, with British Rail, but has subsequently bought out British Rail's share in the project.

The proposed infrastructure fund for London, discussed below, could enable funds to be raised from the private sector for the construction of the east-west CrossRail (Liverpool Street-Paddington) mainline rail link. Whilst ensuring that all business ratepayers likely to benefit from the investment would make an appropriate contribution to the cost, this form of finance would stretch the concept of 'private' finance to the limit. A hitherto standard attribute of private finance is that its payment is voluntary. As a compulsory levy, the use of an infrastructure fund would be closer to the accepted criteria for public funding.

Privatization and rail investment

Rail investment fell sharply in the run-up to privatization. New train orders were reduced to such an extent that the York works of ABB (now Adtranz), where multiple unit trains were built, was forced to close down at the beginning of 1996. Private sector funding completed the Channel Tunnel, although only at the cost of running up a burden of debt and interest it was unable to service. The recent profile of domestic railway infrastructure investment shows a steep fall.

Prior to privatization, the railways underwent a similar investment hiatus to that which the bus industry experienced in the late 1980s and early 1990s. The process of privatizing the operation of railway passenger services is different from that of the bus industry. Bus companies were bought outright from their state or municipal owners, thus incurring substantial debts, which had to be repaid, whereas train operating companies lease rolling stock. However, recovery in the post privatization period will, on some routes, be quicker than it has been in the bus industry, as a result of conditions attached to some of the train operating franchizes. Despite franchize contracts of shorter duration than the life of a train, the companies awarded the Chiltern (Marylebone-Birmingham) and Regional Railways North East (Leeds northern electric services) contracts, for example, are required as a condition of the award to invest in new rolling stock. The Leeds northern example is particularly pertinent, as the lines were only equipped on an interim basis with second hand trains some 35 years old because no one would bear the risk of the investment in the run-up to privatization in the early 1990s.

Criticism has been levelled at franchizes that allow older rolling stock to remain in service, most notably where British Railways Mark 1 designs, with lower collision resistance than more recent stock, are involved (Ford, 1996, pp.430-31).

Tramway competition

Following the passage of the Tramways Act, 1870, tramway construction proceeded only slowly, for a combination of technical and legislative reasons. Horse trams were too slow to offer significant opportunities for expanded catchments and steam attracted only a limited following as a means of power for trams in Britain (Hibbs, 1968, pp.35-36). Electric traction, which brought trams their competitive advantage, only became a feasible proposition during the 1890s and was not widely adopted until after the turn of the century. Tramway development was also deterred by two conditions imposed on tramway operators by the Act. The first of these was contained in section 28 which required tramway companies:

> At their own expense, at all times (to) maintain and keep in good condition and repair, with such materials and in such a manner as the road authority shall direct ... the road whereon any tramway ... is laid ... (and) so much of the road as extends eighteen inches beyond the rails of and on each side of any such tramway.

Tramway owners were obliged by this clause to meet costs in excess of those they imposed on the roads. The beneficiaries included their competitors. The second condition which deterred tramway development was section 43 of the Act which empowered local authorities to take over tramways 21 years after their construction had been authorized, or at seven year intervals thereafter if the power was not taken up within the permitted six month period. This resulted in a lack of investment in system modernization (especially electrification) by private sector tramway owners. Owners were reluctant to invest because they knew that on purchase by a local authority they would only be compensated up to the physical value of the assets, with no allowance for the value of the system as a going concern (in accounting terms, no 'goodwill'). In outer south west London, however, London United Electric Tramways (LUT) developed an extensive electrified network, comprising both former horse lines and new construction, in the knowledge that the lines were safe from local authority takeover for a decade or more (Smeeton, 1994, p.53).

The decade from 1895 to 1905 thus typically saw privately owned horse or steam operated tramways taken over by municipalities, thereby ushering in a century of public ownership of public transport. Electrification usually took place shortly after acquisition. Many local authorities had a dual interest in tramway electrification, as electricity supply authority and owner of the local power station. The combined influences of municipalization, service quality and quantity, and the technical superiority which electric trams acquired over other modes of transport at the time, led to major expansion. The early years of the twentieth century saw tramway route mileage increase from 1,000 to 2,500. In London, trams carried 361 million passengers in 1902, the last year before electrification of the network began, and 812 million in 1913 (Munby, 1978, p.535).

Some authorities, and even private companies such as LUT invested in tramways where the likelihood of making a long term return was slim. Tram services in less densely populated locations, for example some of the outer south west London suburbs served by LUT, and the complete networks in smaller towns and cities such as Darlington, Ipswich and York were among the earliest to be replaced by buses or trolleybuses. The latter enabled local authorities to continue to supply electricity from the municipal power station to the public transport undertaking.

Research by Yearsley (1987, pp.17-34; 1996, pp.3-12) has identified deficiencies in the financial policies of tramway undertakings, which resulted in equipment requiring replacement before it had been fully written down. The outcome was that the operators experienced periods of double debt, for example in respect of track renewals. The reduction of loan periods, typically from 40 to 25 years, gave higher annual repayments than previously. Yearsley expressed concern that the lessons from the tramway era may not have been learned since no renewals fund has been created in respect of Manchester Metrolink or Sheffield Supertram, the government apparently preferring the PTEs to borrow money when it is required.

Underground railways

The initial investment in underground railways in Britain was made by private companies in London, Liverpool and Glasgow; in London, the Metropolitan, Metropolitan District, City and South London and Central London Railways, for example; in Liverpool the Mersey Railway; and in Glasgow the Glasgow District Subway Company.

The mainline railway companies did not generally become owners of deep level underground lines, although London's Waterloo and City line belonged to the London and South Western Railway and the Great Northern Railway was also involved in tube promotion. The Waterloo and City remained with its mainline parent and became part of British Railways in 1948. It was eventually transferred to London Underground (LUL) in 1994 as an adjunct to the Central Line, prior to the privatization of South West Trains (Glover, 1996, p.157). Some 70 years after being built to mainline proportions, with the intention of taking Great Northern Railway suburban trains to the City of London, the former Great Northern and City line eventually became part of the British Rail Great Northern Electric system (now West Anglia Great Northern) in 1976.

Subsequently, ownership of underground railways has moved in different directions in each of the three cities.

London

International finance in London in the early twentieth century   Private investors, notably the controversial Charles Tyson Yerkes, miscalculated in the matter of funding the construction of London's deep level tube lines in the Edwardian era. Yerkes' interest in investing in underground railways in London stemmed from his realization of the potential for interchange with mainline railways afforded by the proposed Charing Cross, Euston and Hampstead line. By 1902, he controlled three further underground railways (District, Great Northern Piccadilly and Brompton [Piccadilly Line] and Baker Street and Waterloo [Bakerloo Line]), which he brought under the unified control of the Underground Electric Railways Company of London (UERL), together with the LUT. The direct cause of Yerkes downfall was over optimistic forecasts of ridership and revenue, which had predicted usage at around double that which actually materialized (Barker and Robbins, 1974, pp.116-17). One outcome of this crisis was the introduction of through fares between lines (Barker and Robbins, 1974, p.143), which provides an interesting precedent for the success of Travelcards in generating additional custom in the 1980s and 1990s. Recovery and further amalgamation took place after the death of Yerkes in 1905. Most notable was the absorption of the London General Omnibus Company (LGOC) into the Underground Group in 1911. As noted in Chapter 2 this Group, under the stewardship of Albert Stanley (later Lord Ashfield), provided the foundation for London Transport.

Yerkes was not the only major investor in London's underground. The Central London Railway, forerunner of today's Central Line, was financed by an international banking syndicate whose principal shareholder was the Exploration Company, a company founded to develop foreign mining interests and backed by Rothschilds (Barker and Robbins, 1974, p.39). A fictional trilogy by Theodore Dreiser captures the atmosphere of the buccaneering financial style epitomized by Yerkes. The three volumes (Dreiser, 1912, 1914 and 1947) follow the career of 'Frank Algernon Cowperwood' from his Philadelphia upbringing, through financial exploits in Chicago, to his final years, most of which were spent in promoting underground railways in London. Cowperwood is a thinly disguised Yerkes. In the words of Barker and Robbins (1974, p.62), Dreiser, a journalist, thoroughly researched the background to his novels, with the result that 'it is difficult at this distance of time to winnow Dreiser's fiction from his facts; but much rings true'.

Neither private nor public finance succeeded fully in providing central London with a comprehensive network of underground railways or tramways. The combined effects of competitive bids by rival financial consortia and parliament's unwillingness to allow every proposal to go ahead, have left gaps in the underground network which still persist. No underground line has ever followed Edgware Road or Park Lane, linked Kensington to Hyde Park Corner, or served Fleet Street and the Strand satisfactorily; each major demand axes for which proposals were rejected in the Edwardian era.

Nor did the London County Council's (LCC) plans for a network of shallow tram subways develop beyond the experimental Kingsway subway (1906-52), which remained unique in Britain. Perhaps future generations may be more grateful that another LCC proposal, for a huge east-west highway on the alignment of Langham Place to Russell Square, and wide enough for four lines of tram tracks, also failed to materialize. Half a century later, proposals for continental-style tram subways in Leeds, similarly failed to be implemented.

The formation of London Transport London's underground railway companies were not included in the railway grouping of 1923, but as explained in Chapter 2 became part of the London Passenger Transport Board (LPTB) on its formation a decade later. Public financial support for expansion of the network began in the late 1920s, under a series of government unemployment relief initiatives in the form of development loan guarantees at preferential rates of interest. Transport projects to benefit included extension and modernization of the London Underground, railway electrification and modernization, the construction of arterial roads, and accelerated programmes of investment in tramway replacement. For example, the Development (Loans, Guarantees and Grants) Act, 1929, facilitated the modernization and extension of the Piccadilly Line (Lee, 1973, p.20).

According to Glover (1996, p.49), 'it was only with the creation of the [London Passenger Transport] Board in 1933 and the institution of financial pooling agreements with the 'Big Four' mainline railways, that it became practicable to consider the needs of London as a whole'. The 1935 'New Works Programme, involving the LPTB, LNER and the Great Western Railway (GWR), arose from this inter-operator co-operation. The programme brought about the extension into north and north east London of the Central and Northern Lines and, eventually in the 1960s, electrification of the Metropolitan Line to Amersham and Chesham. The LNER, which had been unable itself to afford their electrification (Morrison, 1933, p.61) lost its Epping, High Barnet and Edgware branches to the LPTB and gained electrification from Liverpool Street to Shenfield.

The intervention of World War Two prevented completion of the Northern Line project, principally because of the creation of a green belt around the built-up area of London immediately after the war. The proposed, and indeed partly built, Bushey Heath extension would have traversed unpopulated territory, and was therefore logically abandoned. The rationale for failing to finish the substantially prepared sections within London probably owed more to short term financial considerations, rationalized by revised traffic and revenue projections. Crouch End and Muswell Hill, in north London, have been without rail links since closure of the Finsbury Park to Alexandra Palace line, an unfulfilled part of the 1930s Northern Line extension programme, leaving these areas among the few in London solely reliant on buses for their public transport.

The Victoria Line, authorized in 1962, originated as 'Route C' of the Railways (London Plan) Committee, whose report was published in 1949 (Ministry of Transport, 1949, pp.15-16). The line was justified on social cost-benefit criteria, which included a substantial element in respect of benefits to road users, rather than on the basis of a conventional financial appraisal. Although the Victoria Line increased the total supply of transport, much of the supposed benefit to road users has been dissipated through the release of formerly suppressed demand for car travel into actual trips. For such an approach to work, highway capacity management, such as the creation of bus priorities and the introduction of traffic calming, needs to be introduced when a new railway opens, to restrain car use at its previous level.

There is now a political and professional awareness of this factor, for example where 'Red Routes' are introduced, that did not prevail in the 1960s. The aims for the Red Route network laid down by the Secretary of State for Transport are 'to improve the movement of all classes of traffic on the Priority Red Route Network...(to) provide special help for the efficient movement of buses...(and) to do so without encouraging further car commuting into central London' (Traffic Director for London, 1993, para.3.2). The 500 kilometre Red Route network is scheduled to be implemented in full by 1999. The project is government funded; in 1995-96 the grant was £15 million (Traffic Director for London, 1996, p.23).

LUL is now a separate company owned by London Regional Transport, and presently in public ownership although private funding has begun to play a role both in capital projects and operations. The private sector has contributed to the cost of the Jubilee Line extension and GEC Alsthom have a performance-related contract for the provision and maintenance of the Northern Line's new trains for 20 years (Glover, 1996, p. 139). The concept of totally privatizing the London underground as a series of individual line-based companies was floated by Glaister and Travers (1995), but this option did not receive serious consideration. In 1996 privatization was ruled out by the Conservative Government, leaving LUL among the last companies to remain in the once comprehensive publicly owned transport sector.

When seeking private funding for public transport investment it may be difficult to ensure that all beneficiaries meet their share. If, for example, only major property owners or businesses fund a transport development, others who have not contributed may also gain. Whereas British Gas was the main beneficiary from its contribution to the cost of diverting the Jubilee Line extension to serve a new development on its land on the Greenwich peninsula, it will be more complicated to ensure that all West End businesses contribute towards the cost of a privately financed Liverpool Street to Paddington Cross Rail.

Travers and Glaister (1994) have prepared a case for an infrastructure fund for London, to which a 'convincing majority' of business ratepayers would be prepared to contribute through a non-domestic rate levy. This proposal has been endorsed by 'London First' in its London Transport Initiative (1996, p.23). 'London First' brands as 'bizarre' Treasury Ministers' objection to the proposal on the grounds that the rate surcharge would amount to an additional tax and that the additional spending would count as public expenditure.

Liverpool

In Liverpool, the Mersey Railway was the first line in Britain to be converted from steam to electric traction (in 1903, funded by raising new capital) and has subsequently passed via the London, Midland and Scottish Railway and British Rail to Merseyrail Electrics Limited. Unlike the Liverpool Overhead Railway (LOR), discussed below, the Mersey Railway's line has survived to become part of the improved and busy Liverpool 'loop and link' underground railway system, which opened in 1977-78. The loop and link connect formerly separate suburban railways across the city centre, as subsequently achieved in Newcastle by the Tyne and Wear Metro and in Manchester by Metrolink. The 'loop' extends the former Mersey Railway line in a circuit around the city centre, enabling it to perform a distributary function. Trains operate on an out and back basis from the Wirral without the need to reverse or wait in Liverpool. The 'link' joins the northern (Southport) and southern (Hunts Cross) routes across the city centre and provides interchange with the loop.

Planning for the loop and link preceded the formation of the Merseyside Passenger Transport Executive (MPTE). The Mersey Railway Extensions Bill was sponsored by British Rail, with Liverpool City Council acting as trustee for the future conurbation authority. 75% of the funding for the loop and link came in the form of an infrastructure grant from the Ministry of Transport and 25% from the MPTE (Merseyside Passenger Transport Executive and British Rail, 1978, p.13).

A unique British railway, and one whose funding and ownership were also unique, was the LOR, which opened between 1893 and 1896 and closed at the end of 1956. The LOR was promoted, but not actually owned, by the Mersey Docks and Harbour Board (MDHB), a public trust port company, whose membership was dominated by shipowners and merchants (Dyos and Aldcroft,

1971, pp.251-54). The line was electrically operated from the outset 'because of the risk of cinders and sparks falling in the dock area' (Holt, 1978, pp.40-42). In 1903, proposals to link the LOR with the suburban railway services of the Cheshire Lines Committee (CLC), to create a circular route, came to nothing. If the CLC plan had materialized a less isolated LOR might have enjoyed a longer life, but this was not to be. Faced with the need to spend £2 million to renew the decking of the overhead structure in the 1950s, the company tried, without success, to persuade Liverpool City Council and the line's original sponsor, MDHB, to take it over. Although still carrying some 10 million passengers annually, it was forced to close, over a decade before the creation of the MPTE, with the power to support public transport (Transport Act, 1968, section 10). Perhaps the LOR's premature demise saved it from a slow death by cuts, as many of the docks it serviced shut, and inner city population density declined. Or maybe it could have become an engine for inner city regeneration, meeting the needs of the reviving Liverpool waterfront, in the manner of the similarly elevated Docklands Light Railway in London.

Glasgow

The Glasgow underground came into Glasgow Corporation ownership in the 1920s and was transferred to Strathclyde (originally Greater Glasgow) Passenger Transport Executive (SPTE) in 1973. The underground was comprehensively reconstructed with the aid of finance from SPTE and reopened in 1980. In 1979 the east-west suburban link via Glasgow Central Low Level, which had closed to passengers in 1964, was electrified and reopened as the Argyle Line (Thomas, 1971, revised by Paterson, 1984, p.247). Finance for the project came from the PTE.

Buses

Early investment in bus operation was achieved through a mixture of private firms and local authorities. The industry was profitable overall, a situation which continued into the 1960s. Where losses did occur, cross-subsidy was available, especially after the introduction of Road Service Licensing under the Road Traffic Act, 1930. This had the effect of protecting incumbent operators from excessive competition, with the result that comfortable profits could be earned. Municipal operators also effected cross-subsidy from buses and trams to other corporation services or to the relief of rates. It was only when falling profits began to leave insufficient resources for cross-subsidy that the need for external financial support arose. In an interview given in 1988, A.N.Todd, the National Bus Company's (NBC) chairman from 1969 to 1971, highlighted the inequity of cross-subsidization on a wide scale. Todd observed that NBC had attempted during his period of chairmanship to attack the subsidization of loss-making companies in the prosperous south of England from profits earned by subsidiaries in the north (Birks et. al., 1990, p.261).

The rebate of duty on diesel fuel used for local (stage carriage) bus services was introduced in 1964 under the Finance Act, 1965, and an increase in the maximum amount payable sanctioned by the Transport Act, 1968 (section 33). From 1974 to 1993 the full duty was rebated. Thereafter, the level of rebate was frozen at its 1993 level, but the level of duty was increased by 5% or more annually, as part of a measure intended to constrain the demand for car travel. This has had the perverse effect of increasing bus operating costs relative to the cost of car travel and cost the bus industry £165 million between 1993 and 1996 (Confederation of Passenger Transport, UK, 1996, para.2).

New bus grant was introduced under the Transport Act, 1968, section 32. Unfortunately this grant was tied to the purchase of approved types of vehicle, with the specific objective of increasing the bus industry's cost effectiveness by encouraging the move towards one person operation. Fare structures and methods of fare collection should have been radically revised at the same time, but were not. The result was increases in boarding time per passenger, typically from around 1-1.5 seconds to 3 or more, necessitating extra running time, especially on busy services. Although costs were reduced, the saving was eroded by the payment of higher wages for one person bus drivers and by the reduction in bus speeds. Despite the almost universal introduction of electronic ticket machines, the dilemma of services which are cheaper to operate, but slower and thus less attractive to passengers, has still not been fully resolved. The original 25% grant was raised to 50% in 1971 and phased out during the mid 1980s (Birks et. al., 1990, p.107, p.273), by which time the objective of converting services to one person operation had been completed in most parts of the country. Central London remained the main centre of bus operation with two person crews.

At first, attempts were made to earmark financial support to bus companies for specific purposes; rural services, railway replacement buses and grants determined and paid by PTEs. Until 1986, grants could be paid in respect of low fares as well as of higher service levels.

General revenue support was first paid to the bus industry in the early 1970s (£10 million in 1972) and reached a peak of £600 million in 1984-85 (approximately £1.1 billion at 1996 prices), before falling to £280 million (almost £300m at 1996 prices) in 1994-95, a reduction of 71% (Transport Statistics Great Britain, 1992, p.34; 1995, p.30). As noted in Chapter 3, the Transport Act, 1968, gave Passenger Transport Authorities (PTAs) widespread powers and duties in respect of local public transport, including the ownership of bus undertakings and the right to precept constituent local authorities. The

Local Government Act, 1972, created metropolitan counties (section 1) which covered the existing PTAs and led to the creation of additional PTEs (section 202). Section 203 gave non-metropolitan counties the duty to co-ordinate and the power to support public transport.

A separate structure applies to the provision of concessionary fares for elderly and disabled people. Such payments are considered to represent the payment of fares on behalf of particular categories of bus user, rather than a subsidy to the operator of the service. Levels of concession offered have been reduced in many locations, typically in the metropolitan areas by substituting a flat fare for free travel, although in London free bus, underground and rail travel has been retained. In a few areas, other concessionary categories are offered, for example to the unemployed or to 'job seekers'. Reduced fares for children, traditionally half, but often now a smaller reduction, are normally offered on a commercial basis by public transport operators. Some operators restrict the concession to ensure that children cannot travel at cheap fares during the peak, when the marginal cost of catering for them is highest.

Since 1986, the Transport Act, 1985, has brought in a fundamentally different approach to the payment of financial support to bus operators. Apart from a broad desire to reduce public spending, the government was concerned that subsidies were 'leaking' into inefficiency. Whilst outdated industrial and managerial practices and bureaucratic organizational structures have, rightly, been abolished, there have also been signs that cost cutting has gone too far. The market, at a time of high unemployment has allowed bus drivers' wage levels to fall from a position above the average for all manual workers to a position below the average. This is neither good for the individuals, nor, in the longer term as employment picks up, for service quality. Similarly, cost cutting by reducing staffing levels has resulted in company structures so lean in some cases that attention to 'soft' elements such as marketing, planning and even vehicle presentation has been deficient. Subsidies for low fares are no longer legal under the 1985 legislation. Support for service levels is restricted to routes or timings which operators have not been willing to provide commercially, that is by registering a local bus service and operating it without recourse to public support. Where a local authority or PTE wishes to purchase anything extra, it must invite tenders for the operation of the additional services. Nationally, around 15% of bus service kilometres are procured in this way, with 85% run commercially. It is often evening and weekend services which cannot be provided commercially. A form of service enhancement which it is difficult to cater for is increased frequencies on a commercial route. Running additional, tendered mileage in parallel to an existing commercial service would be expected to generate some extra ridership, but less than proportionally to the increase in service level. It would thus tend to reduce the commercial service's profitability.

Bus industry investment post deregulation

During the decade since bus deregulation, a period during which surplus funds have been dedicated to repaying bank loans incurred at privatization, rather than to investment in tangible assets, investment is beginning to pick up again. Bus companies, especially the larger ones, have begun to reinvest in new buses, with the result that the average age of their fleets is once again falling. The incentives provided by stronger emission controls and the availability of fundamentally new types of vehicle, such as ultra low floor and gas powered, will each encourage investment, as companies seek their place at the forefront of developments. Big cities have dominated in the first deliveries of low floor buses (London, Newcastle and Manchester) and gas buses (Southampton, West Midlands).

New approaches to running buses are resulting in an increasing number of park and ride services, most of which involve substantial public/private partnerships. Synergy between landowners, superstore operators, local authorities and bus companies is reaping considerable benefits in terms of freeing town and city centres from excesses of traffic (Confederation of Passenger Transport, UK, 1995). Bus-based park and ride features most commonly in medium-sized, historic cities, such as Oxford, Cambridge and York. In the largest cities, rail-based park and ride is more usual. In most locations, park and ride bus services are run on contract to the local authority, although in Oxford competing, commercially operated services are run. Local authorities prefer the degree of control which a contract service gives them, compared to a commercial service.

A combination of private finance deals and a slightly relaxed attitude by the government is seeing increased investment in new light rail systems. Both the Croydon and West Midlands systems were authorized in 1994-95, whereas previously only one system at a time had been sanctioned. Construction of the Midland Metro began in 1995. The contract to design, build, operate and maintain Croydon Tramlink was awarded to the private sector consortium Tramtrack Croydon in 1996, with the government contributing £125 million out of the £200 million cost of the project.

The funding of roads

With roads, a different financial and organizational regime has traditionally applied. Infrastructure has always been provided separately from operations, with financial responsibility split between central and local government. A key component of the financial structure, whose legacy is still with us, was the Road

Fund. The objective of the this fund was to receive income from motorists, which was fed back into road construction and maintenance. Total road costs were met jointly by the Road Fund, together with a contribution from the rates. In 1930, the balance was two thirds ratepayers, one third Road Fund; the Royal Commission on Transport (1931, para.227) considered that these proportions should be reversed (para.249). The new model of railway ownership referred to earlier with separate infrastructure, rolling stock leasing and train operating elements brings rail closer to road in terms of its structure than it has been since the 1830s, when the practice of attaching private carriages to trains enjoyed a shortlived existence.

Although the example relates to long distance travel more than to urban areas, it is nevertheless of interest in the late twentieth century to cite the 1930 Royal Commission's praise for the eighteenth century turnpike trusts which, whilst 'never popular', because tolls were levied, nevertheless 'resulted in an ever increasing mileage of good highway throughout the country' (para.25). The Royal Commission 'strongly opposed' any idea of a network of tolled motorways [was this the first time the term 'motorway' was used?] constructed by private enterprize (para.222-23). The issue of charging for road use has still not been resolved satisfactorily, with a few schemes for privately funded roads in the pipeline, but with no real confidence that this is the way forward. Tolling a small proportion of the network still raises the spectre of diverting traffic back into areas from which it has been expelled, just as it did in 1930. Urban road pricing has recently been deferred again in a study on London. The need to get users to meet the costs they impose - which, it is becoming increasingly clear, must include externalities - is obvious, but the will to adopt a particular method of doing it continues to be absent. The issue is fundamentally the political one of how to persuade people to accept that they must pay for something they have been used to getting for free at the point of consumption.

Conclusion

The balance of issues has changed but little over the past century or more. Concern still centres fundamentally on whether transport is a public good, provided for and to be paid for by everyone, or whether it should be operated entirely as a business, paid for as it is consumed.

Traffic congestion and environmental pollution were concerns in the 1890s, and are still with us. The century has seen the rise and partial fall of collective transport, but with, at the time of writing, an increasing acceptance that individual mechanized transport is too inefficient in its use of land and road space to be allowed to maintain its present dominance in urban areas; dominance, that is, in terms of highway and parking provision, but not in terms of modal share of the market. Public transport's renaissance, we constantly hear, is 'just around the corner'; but the corner is taking a very long time to turn.

There are encouraging signs, however, that the situation is changing in the locations where traffic problems have reached the greatest proportions. Quality bus corridors are being created in several cities, including guided busway projects. Several light rail systems are in operation, with more under construction and planned, although the warning signs regarding their cost must be heeded, and the number of places where they will be built will be smaller than some would wish. The commercialization of heavy railways may enable them to increase service quantity and quality more cheaply than would have been imagined only a few years ago.

Above all, it is to be hoped that consistent and equitable evaluation systems and financial structures will be implemented to ensure that Britain's cities in the twenty first century are pleasant, clean, quiet and safe places in which to live, work and play.

The opinions expressed in this paper are those of the author and do not necessarily reflect the policies of his employer.

References

Barker, T. and Robbins, M. (1963), *A History of London Transport: Volume 1 - The Nineteeth Century*, Allen & Unwin, London.

Barker, T. and Robbins, M. (1974), *A History of London Transport: Volume 2 - The Twentieth century to 1970*, Allen & Unwin, London.

Birks, J. (1990), *National Bus Company 1968-89: a commemorative volume,* Transport Publishing Company, Glossop.

Bosley, P. (1990), *Light Railways in England and Wales*, Manchester University Press, Manchester.

British Railways Board, (1963), *Reshaping of British Railways*, HMSO, London.

British Railways Board, (1992), *Annual Report and Accounts 1991-92*, HMSO, London.

British Transport Commission, (1961), *Annual Report and Accounts 1960*, HMSO, London.

Confederation of Passenger Transport, UK, (1995), *Park and ride bus services: opportunities and prospects,* (unpublished), London.

Confederation of Passenger Transport UK, (1996), *Fuel duty rebate: submission to HM Treasury,* London.

Connor, J. (1996), 'The Greenwich Park branch' in *London Railway*

Record, Colchester.

Dreiser, T. (1912), *The Financier*, Harper, New York.

Dreiser, T. (1914), *The Titan*, John Lane, New York.

Dreiser, T. (1947), *The Stoic*, Doubleday, New York.

Dyos, H. and Aldcroft, D. (1971), *British Transport*, Leicester University Press, Leicester.

Ford, R. (1996), 'Mark 1 spectre at franchizing feast' in *Modern Railways*, Ian Allan, London.

Glaister, S. and Travers, T. (1995), *Liberate the tube, Policy Study No.141*, Centre for Policy Studies, London.

Glover, J. (1996), *London's Underground*, 8th edition, Ian Allan, London.

Hibbs, J. (1968), *A History of British Bus Services*, David & Charles, Newton Abbot.

Holt, G. (1978), A *Regional History of the Railways of Great Britain, Volume 10: The North West*, David & Charles, Newton Abbot.

Jackson, A. (1973), *Semi-Detached London: Suburban Development, Life and Transport 1900-39*, Allen & Unwin, London.

Jackson, A. (1978), *London's Local Railways*, David & Charles, Newton Abbot.

Lee, C. (1973), *The Piccadilly Line: a brief history*, London Transport, London.

London First Transport Initiative for the London Pride Partnership, (1996), *London's action programme for transport: 1996-2010*, London.

Merseyside Passenger Transport Executive and British Rail, (1978), *The story of Merseyrail*, Liverpool.

Morrison, H. (1933), *Socialization and Transport: The Organization of Socialized Industries with particular reference to the London Passenger Transport Board*, Constable, London.

Munby, D. (1978), *Inland Transport Statistics Great Britain 1900-70*, Clarendon Press, Oxford.

Royal Commission on Transport, (1931), *Final Report*, HMSO, London.

Savage, C. (1966), *An Economic History of Transport*, Hutchinson, London.

Sherrington, C. (1969), *A Hundred Years of Inland Transport*, Frank Cass, London.

Smeeton, C. (1994), *The London United Tramways: Volume 1 - Origins to 1912*, LRTA/TLRS, London.

Transport, Department of (1983), *Railway Finances*, HMSO, London.

Transport, Department of (1995), *Transport Statistics Great Britain, 1995 Edition*, HMSO, London.

Transport, Ministry of (1949), *British Transport Commission London Plan Working Party, Report to the Minister of Transport*, HMSO, London.

Thomas, J. revised by Paterson, A.J.S. (1984), *A Regional History of the*

Railways of Great Britain, Volume 6: Scotland - the Lowlands and the Borders, David & Charles, Newton Abbot.

Tomlinson, W. (1987), *The North Eastern Railway: its rise and development*, David & Charles, Newton Abbot.

Traffic Director for London, (1993), *Network Plan Consultation Document*, London.

Traffic Director for London, (1996), *Annual Report 1995-96*, London.

Travers, T. and Glaister, S. (1994), *An Infrastructure Fund for London*, Greater London Group, London School of Economics, London.

Yearsley, I. (1987), 'Previously unexplored aspects of London's tramway finances' in Higginson, M. (ed.) *Tramway London: background to the abandonment of London's trams 1931-52*, Light Rail Transit Association, Birkbeck College & London Transport Museum, London.

Yearsley, I. (1996), 'Light rail: who pays?' in *Proceedings*, Chartered Institute of Transport, Volume 5, No.2, London.

6 Influences in transport development

James McConville

Introduction

The last two decades have witnessed a formidable movement towards deregulation and greater market flexibility in the pursuit of increased levels of competition. Pressure for these changes emanated from the impact of inflation; technological and organizational innovation; profound changes in the sphere of economic thought; and, finally, the government's desire to realign its activities and expenditures. The result has been a substantial undermining of the long held notion that many of the industry's problems sprang from the difficulties generated by unrestrained competition. This view was particularly influential in transport where the desire to ensure certainty of supply prompted the establishment of elaborate regulatory systems with controls on prices and limits on entry, together with an emphasis on user safety and security of employment.

The period has also seen the attitude of policy makers to competition undergo radical change. From constituting the problem it was increasingly portrayed as a solution. The outcome was substantial, piece-meal deregulation in an attempt to exploit the benefits of competition. However, viewing the situation in the late 1990s, it is apparent that the drive for greater deregulation and increased competition has lost some of its vigour. New challenges, including chronic urban traffic congestion, environmental pollution, uncertainty of service provision, adverse employment prospects and the impact of increasing

European integration are once again raising questions about the benefits of unrestrained competition and deregulation.

The transformation of competition from being the inherent problem to becoming the solution owed much to the 'public choice' school of economic thought which developed in the 1970s. This coincided with the state, political and other groups changing their perception of the role and function of transport in the late twentieth century. Increasingly it had been regarded as a combination of public utility and welfare service. Recently, however, greater pressure has been exerted to impel transport into a competitive self regulating market structure. This chapter aims to analyze the interplay of comparatively consistent economic and technological criteria in the face of rapidly changing social, commercial and economic conditions. It concludes that, notwithstanding the slackening in the drive for greater competition, a metamorphosis has taken place in the British transport industry whereby the state's role as a provider has been replaced by the state as a regulator.

Background

Although it was nascent in the prewar period it was the arrival of postwar prosperity which really established the phenomenon of 'consumerism'. This is an obsessive concern with the private acquisition of consumer goods, not least the motor car. The car rapidly and profoundly influenced the individual's idea of movement; a high value being placed on the possession of flexible, personal mobility. The car is now perceived as an integral part of individual liberty, with the right to drive and park virtually at will, a 'right' inconceivable to poorer and earlier societies. To this end both society and its individual members deploy larger and larger amounts of their economic resources to satisfy the needs of individual transport. In 1994-95, for example, the average household spent 15% of weekly expenditure, (i.e. £43), on travel and 80% of this sum (i.e. £36) on the purchase and running of a motor vehicle.

Society has become committed to the car, or more broadly, road transport. This has been a cardinal factor in the re-emergence at intervals of what has been termed the 'transport problem' and particularly the 'urban transport problem'. Britain is at present experiencing such a period when there is a general perception of a 'transport problem'. There were similar shifts two decades ago and during the inter-war years. The perception of a 'transport problem' is one of congestion, peaking, loss of social welfare, wasteful use of economic resources and environmental despoliation. This concern has much to do with the characteristics of the industry from the consumer's point of view.

The base role of transport is to provide a system for the conveyance of people and goods. Here the discussion is limited to public rather than individual or private conveyance. Its most notable characteristic is that nobody wants it. It is not demanded in its own right but rather as a means to an end; a service or facilitator of other objectives. Here a tiny minority of transport 'enthusiasts' or 'eccentrics' are being ignored. Passengers wish to be in another location. Freight needs to be transferred elsewhere. To be at some other destination is the central objective. Related to this is the perception of transport efficiency. This is seen in terms of time, specifically a mixture of speed and the cost of delivery; the fastest time at the least cost is equated with the most efficient method of movement. Virtually all transport investment is to shorten the 'time costs'. Transport is therefore seen by passengers as a negative commodity. Unlike most other productive commodities, transport cannot be stored and therefore if it is not utilized when it is available it is lost. This is particularly important when regular services are involved. Cheap day tickets, and other inducements to travel are simply to add, often anything, to revenue because the suppliers costs have already been established once the vehicle is in service. Further, the consumption of transport takes place within the provider's, that is the supplier's, property. Unlike most other commodities which are purchased and consumed elsewhere, the consumption of transport is virtually by definition within its facilities. It follows that consumers become particularly exercized by queuing, delays and congestion. These are part of the negative externalities of the industry which arise because:

Transport is an engineering industry carried on not privately within the walls of a factory, but in public places where people are living, working, shopping and going about their daily business. The noise, smell, danger and other unpleasant features of large, fast-moving machinery are brought close to people, with potentially disastrous consequences for the human environment. These are obviously external costs associated with most forms of transport engineering. Such massive industries utilize an enormous amount of economic resources (Thomson, 1974, p.46).

In industrially rich countries, i.e. members of the Organization for Economic Co-operation and Development (OECD), consumer expenditure on transport represents between 4% and 9% of gross national product. In the United Kingdom it accounts for approximately 17% of total consumption expenditure and some 2% to 5% of employment. The transport sector in this country is fortunate in possessing a massive inheritance of economic assets. These are fixed assets, mainly infrastructure, with extremely long lives. For example, Roman roads, ports built in the Middle Ages and early nineteenth century railways are incorporated into our transport heritage and present day activity. All very expensive in terms of economic resources, having extremely long economic lives but with a negative property having no alternative use. These cannot be changed apart from in the most exceptional circumstances. A railway tunnel has no other use than the one for which it was constructed. Generally, the technology becomes obsolete. Canals and much of the waterway system are an example of this obsolescence which the leisure industry has been unable to fully absorb. Obviously to move a single passenger or tonne of freight by rail between Liverpool and Newcastle requires the basic facilities of a railway. However, once established vast numbers of passengers and quantities of freight can be handled. Such enormous consumption of resources combined with vast economies of large scale production raise historical apprehensions about 'natural monopoly' and the liberal fear of the impact of market failure. Railways are a classic case. Hence the government's involvement has often meant control, ownership or simply massive investment on economic, social and, occasionally, strategic grounds.

In contrast to the infrastructure of transport the mobile unit, the vehicle, i.e. the car, the van, the lorry, etc., differ substantially. The vehicle has a limited physical life, something in the region of 15 years. Most have a multiplicity of uses, for example a road freight vehicle can perform a number of functions. There is a distinct limitation on the economies of scale which is influenced by two factors, the level of demand which is generally thinly distributed over time and geographical space. Buses are a good example from the passengers' point of view, they want a maximum flexibility of time and place, and the provider's ideal would be a very large capacity bus with predictable uniformity. However, few people want a bus from exactly the same place to another specific place at exactly the same time. Hence attempts at some optimum level of operation imply the acceptance of congestion and overcrowding. Combined with this, transport infrastructure often limits economies of scale, as with urban buses which are specifically limited by their ability to turn city street corners.

These attributes make the government's attitude towards vehicles fundamentally different from that of infrastructure. In the case of vehicles they concentrate on social and safety procedures, particularly using licence systems which also present them with fiscal opportunities. These are the basic characteristics of transport which underpin the next stage of this discussion that of analyzing the period since World War Two.

A period of expansion

For some 30 years following the war there was a high level of employment and investment, increased growth and expanding industry and innovation. All of this had an enormous impact on the transport system; passenger miles travelled more than doubling in this period. The car, as already noted, is of prime importance here for in Britain it did not gain a 'social market depth' until some time after World War Two unlike in other industrial countries, for example the United States. In the early 1950s, only 13% of households owned a car and only 1% had two cars. By 1994-95 45% of households owned one car, 19% two cars and 4% three or more cars. It follows that at the beginning of the period 80% of households had no car, whereas today this figure is 31%. Hence by the mid 1990s approximately one third of households did not have regular use of a car and were therefore constrained to travel by some form of public transport or walk. In this period cars created massive growth in personal mobility, a trend having a considerable impact on shopping, leisure, education and other cultural activities. These public and private services are increasingly being concentrated in a declining number of very large units; massive out of town shopping centres are an obvious example. The development of this trend was highlighted in the late 1970s. As Lee observed:

A paradox of the development of the last twenty years has been the increased mobility of a rising proportion of the population, accompanied by the decreased mobility of those without access to private cars. The size of the minority experiencing this problem may diminish in the future, but the severity of the problem could considerably increase in urban as well as rural areas (Lee, 1977, p.26).

Table 1.1 illustrates the expansion of passenger transport during the last four decades in terms of billion passenger kilometres (b.p.k.). It also highlights the car's underpinning of a substantial increase in personal mobility that has produced a fundamental change in people's lifestyles. The distance passengers travelled increased by some threefold during the period from 219 to 689 b.p.k. Within this expansion the car and van mode increased some tenfold. In the process of doing this it appreciated from 27% to nearly 90% of this vastly expanded total. Much of this increase, as with other modes, occurred during the first 20 years of the period. By 1974 this mode constituted some 76% of the total. Such a massive increase put other modes under considerable pressure. The cycle, both motor and pedal, constituted 14% in 1952 but had contracted to a mere 2% by 1974. During the last two decades, apart from some increase in motor cycling during the early 1980s, due perhaps to the increase in the price of fuel, it remained constant at 2% with a marginal increase in real terms to 10 b.p.k. Buses and coaches experienced the most dramatic collapse in passenger kilometres from being the prime mover with some 42% to a mere 6% (in real terms down to 14 b.p.k.). This contraction was similarly largely concentrated in the first half of the period. Roads in general increased from 82% to 94% of b.p.k. Railways in real terms remained relatively constant during the period, with 39 b.p.k. initially and latterly at 37 b.p.k., oscillating between a high of 41 b.p.k. and a low of 31 b.p.k. In a period of massive expansion in passenger movement this of course meant a percentage contraction from 18% to 5%.

Table 1.1
Great Britain passenger transport by mode 1952-94
(% and total b.p.k.)

	1952	1964	1974	1984	1994
Buses and coaches	42	21	14	9	6
Cars and vans	27	63	76	81	87
Cycles[1]	14	4	2	3	2
All road	82	89	91	93	94
Rail	18	11	8	7	5
Total (b.p.k.)[2]	219	340	441	534	689

Notes
1. Cycles = motor and pedal
2. Air less than 1% throughout period

Source: Transport Statistics Great Britain, HMSO, London, (Selected Years)

Engineering model

When examining the impact of these developments on governments it is often overlooked that the perception of the transport problem is influenced (or perhaps created) by researchers, sectional and, in particular, industrial pressure groups. Their perceptions and, until recently, their preferred solutions to the problem, has been one of engineering efficiency through the creation of an infrastructure to allow or ensure the free flow of expanding traffic. Such engineering solutions are dependent obviously on some blueprint, in other words planning. Hence the response was one of increased road construction and high levels of maintenance and this was largely successful in the short run. This is logical for practical decisions are generally a response to specific short term pressures. An example of planning was the ledgendary report *'Reshaping of British Railways'* (BRB, 1963) and the introduction of motorways. Table 1.2 indicates the additional kilometres of roads and motorways in Great Britain during the period 1952-94. It illustrates the additional investment in roads which constituted in the selected period on average a little under 20,000 miles between a 5% and 7% increase. The exception being 1964 and 1974.

Table 1.2
Great Britain additional road lengths 1952-94
(kilometres)

Motorways	Total (additions)	(included)
1952-64	21,885 (7.3%)	480
1964-74	7,581 (2.3%)	1,389
1974-84	18,553 (5.6%)	917
1984-94	17,377 (4.9%)	382

Source: Transport Statistics Great Britain,, HMSO, London, (Selected Years)

Although this era has been termed the engineering period, the assumption here is that the emergence of motorways has had an important impact with the peak in expansion being during the middle of the two decades. In the decade 1984 to 1994 motorway expansion fell to an addition of a mere 382 kilometres. This can be explained from two points of view. The network was established and needed only marginal additions. Second, and more understandably, the vast resources required for motorway construction had come increasingly under the government's expenditure scrutiny. Some elements of the engineering solution still persist but do not hold the sway they did in the early postwar years, largely because of the emergence of economic resource allocation problems. A persistent problem within the economy is that the resource cost of the transport system is increased. It thus became questioned specifically from the point of view of the state and subsidization and its perceived failure to solve the contemporary transport problem.

The underlying assumption of the engineering solution was accepted by governments of all shades during much of the period under discussion but within these parameters there were substantial differences in policy. As Savage pointed out:

Changes in political complexion of successive governments have far from been incidental matters in the postwar history of British transport. Deep difference in political and economic philosophy between the two main parties profoundly influence the organization and successive organization of transport in the two postwar decades (Savage, 1952, p.173).

Throughout this period the philosophic core lay in the different attitudes to private and public ownership. This related closely to how competition was viewed. During much of the immediate postwar period, differences can be observed in the language used. Initially the discussion was on integration and co-ordination. The idea of competition came significantly later. The immediate postwar Labour Government's Transport Act, 1947, saw integration through complete technological monopoly with it being achieved presumably only by common ownership, that is state ownership. The Conservative Government's Transport Act, 1953, replaced the word 'integration' with 'co-ordination'. This owed much to the views of the Royal Commission of 1930.[1] Attempts were made through anti-competitive legislation, to co-ordinate working and development of the available means of transport in Britain. In the 1950s co-ordination was to be achieved through the interplay of 'natural competition' and other competitive forces. There was some denationalization but far from total. All transport modes or sectors were to exist separately to encourage healthy rivalry between them. This largely remained the philosophy through the 1960s and early 1970s. Although the Labour Government's Transport Act, 1968, returned some elements of planning, and also introduced some public welfare elements with a social grant service, there was no return to the earlier policy.

In the context of the nationalized industries, where transport was a major sector, change had begun earlier. In 1961 financial responsibilities were tightened in a number of ways. Most important was the innovation of financial targets. This did little to modify the stringent criticisms of the nationalized industries' general progress. Further action was taken in 1967 with the adoption of investment and pricing procedures, including marginal cost pricing and net present value procedures, in investment decisions. Of far greater importance was the oil crisis of the 1970s and the financial difficulties which the Labour Government found itself in during the later years of that decade. The announcement of an approach to and later a loan agreement with, the International Monetary Fund being conditional on deep cuts in public expenditure was of more than symbolic importance for the economy in general and the transport industry in particular. It was against this background that the primacy of the engineering solution to the transport problem began to recede, as one commentator observed:

It now seems a generation ago that the most conspicuous activity on the transport scene was the pouring of miles of new concrete and the bubbling of asphalt as roads were added to roads in praise of the great god, mobility.

Since then, other idols have displaced it from the centre of transport policy. We now realize anyway, that in cities, where most of the traffic is, the greatest contribution to mobility is through traffic management not road building - through better signalling systems, one way streets and the like (Foster, 1976, p.17).

Here then is a much diluted engineering solution based on management regulatory systems. At the same time the Labour Government's Consultative Document of 1976 under the title 'Transport Policy' pointed out the failure of the 1968 Act as it had not served to create a proper framework for the co-ordination of transport policy both at national and local level.[2] Whilst not directly challenging car ownership it was however conscious of its serious impact on public transport. Although noting the emerging concern for the environment it conceded that the quality of life had improved dramatically. Further, it made the point that the energy crisis of 1973 had generated concern regarding the growing scarcity of fossil fuels. Finally, the primacy of public expenditure in all of this was made absolutely explicit. Thus overshadowing all these developments was the prevailing public expenditure situation. On the one hand public spending on transport had increased rapidly but the government now decided to limit the growth of public expenditure in the interest of higher exports and investments generally. This required, first, a reappraisal of the priorities to be accorded to transport as against the competing claims of housing, education and the social services, and, second, a ruthless re-examination of transport expenditure to ensure that it was actually (which many critics doubted) achieving the social and economic ends in view.[3] Such views have stimulated claims that the fundamental changes of the last two decades were inaugurated by the Wilson and Callaghan governments of the mid and late 1970s. Ball and Cloake, for example, have argued that:

The Labour administration of 1974-79 ... can with hindsight be said to have cleared the way for privatization policies, accepting (the) strictures of the International Monetary Fund in return for substantial loans, and increasing controls over nationalized industry spending. During this period, therefore, certain commitments to social welfare spending were reduced or jettisoned. Despite this Mrs Thatcher's Conservative Government has been credited with developing this approach in a major way (Bell and Cloake, 1990, p.5).

Privatization and deregulation

The Conservative Governments' motives for what has been loosely termed privatization (or for that matter deregulation) was one of dispensing with nationalized industries. Although it could merely be based on a purely ideological argument which sees ownership as central to efficiency and private ownership in this context eminently preferable, what cannot be overlooked is the attractiveness of these sales in terms of lowering the Public Sector Borrowing Requirement (PSBR). This was part of the government's growing concern with the increasingly high levels of public expenditure on transport. During the last 20 years this was the driving force impelling changes in the industries' elaborate system of regulation, although some account was taken of technological and structural changes. All of which was interwoven with an acceptance of the objectives set by business managers as the paramount economic and social purpose. The assumption has been that these objectives coincide with the aims of society at large. That is to say economic growth, material acquisition and consumerism, constitute the central purpose of individual and national life. The result has undermined the previous notion that many of the industries' problems were generated by unrestrained competition and that transport has a social as well as an economic role. This latter view, as argued, was particularly influential in transport. These policy changes occurred in the late 1970s and early 1980s when the attitude to competition in transport underwent a metamorphosis. From constituting the problem it was now portrayed as the solution. The outcome was a plethora of deregulations, re-regulations and privatizations in an attempt to exploit the benefits of competition.[4]

Transport was at the forefront of these trends with the selling off of long distance bus services and local services with the exception of London, as well as of what was seen as British Rail's non-core activities, for example their hotels and ferries. By the mid 1980s two distinct trends had emerged which created a dichotomy in policy making. In political terms there has been increased pressure for deregulation and privatization. This was almost wholly related to the mobile unit, the vehicle. The infrastructure has proved much more difficult to deregulate or change, with the exception of ports. Railways until quite recently highlighted this difficulty and the road network appears particularly intractable. This is not to say there were no private ventures or funding. The Skye Bridge was opened in 1995 and there is a proposal for a Birmingham North relief road which, if it comes to fruition, would become the first privately funded and operated motorway in Britain. Meanwhile, the other trend is the public concern focusing on social and economic welfare, particularly in the context of the environment. A concern creating the need for increased rather than less control. Both these trends underpin the idea of a transport problem and are increasing the involvement of the state in the transport industry. Privatization with deregulation have been the means by which the state, in whatever form, has retreated from being the provider of transport. But following the logic of the 'public choice' school it has become a supervisor and regulator of the industry with increased emphasis on the issues of efficiency and safety. In particular the individual consumer cannot, it is argued, be left to the mercy of oligopolistic providers.

Conclusions

Transport has an extremely high profile, forcing its internal and external activities into the public arena, hence the high level of state interest and intervention. Such intervention relates closely to the structure of the industry where regulation impacts on the mobile unit and investment on the infrastructure which consumes an inordinate amount of economic resources and highlights the inadequacies of the market to supply sufficient general finance. Such a dynamic and intricate industrial structure has and will continue to throw up complex problems. How these should be dealt with has radically altered in the period under consideration. In the postwar years the accepted solution to the transport problem was what has been termed the engineering model. By the mid 1970s the totality of government policy was undergoing a change as the primacy of the engineering model was being replaced by a more complex set of considerations. The main impetus for these considerations emanated from the growing anxiety about increasing levels of public expenditure on transport. This concern did not initially bring forward a competitive solution. As the Consultative Document 'Transport Policy' of 1976 pointed out '... it does not follow that maximum efficiency is the only goal, and still less that the provision of transport can be left to be determined by market forces. For there are other equally important objectives'.[5] The pressure to curtail public expenditure increased during the 1980s when the government's primary objective became control of money supply and the cutting of direct taxation. To this end privatization and deregulation have been the basic methods used by the state in its rejection of its role as a provider of transport. In this movement to a competitive solution there was, virtually by definition, a repudiation of any precise objectives or strategy which could be termed transport policy. In the short term what materialized was a fragmented approach based on individual projects often attracting substantial amounts of private funding (the Channel Tunnel being the prime example). It could be termed an itemized engineering solution. In the longer term the solution to pollution and congestion will prove exceedingly difficult without some coherent strategy or plan. For, following the logic of the 'public choice' school, the state has become a supervisor and regulator of the industry, particularly since a tension has emerged between the competitive solution and heightening public concern with social and economic welfare, especially in the context of the environment. Further, the individual consumer of public transport is obviously vulnerable to what has become an increasingly oligopolistic or even monopolistic sector of the economy. Thus the metamorphosis in the structure and commercial organization of transport has also entailed a fundamental change in the state's role from being a provider of transport to being increasingly a guardian of those who consume it.

Notes

1   Cmnds. 3365, 3416, 3751 (1930), Report to the Royal Commission on Transport, HMSO, London.
2   Department of the Environment, (1976), *Transport Policy : A Consultative Document, Vol. 1*, HMSO, London.
3   op. cit., para. 1.7-1.8.
4   For a recent discussion of these issues see McConville, J. (ed.) (1997), *Transport Regulation Matters*, Cassell, London.
5   op. cit., para. 13.2.

References

Bell, P. and Cloake, D. (1990), *Deregulation and Transport Forces in a Modern World*, Fulton, London.
British Railways Board, (1963), *Reshaping of British Railways*, HMSO, London.
Foster, C. (1976), 'Transport Policy Revised' in *Socialist Commentary*, London.
Lee, N. (1977), 'A review of Current Transport Policies and Objectives in Britain', *Transport Review, No. 113*.
Savage, C. (1952), *An Economic History of British Transport*, Heinemann, London.
Thomson, J. (1974), *Modern Transport*, Penguin, Harmondsworth.

7 Metropolitan transport policy in the 20th and 21st centuries

William Tyson

Development of transport policy in the twentieth century

It is only in the period since 1968 that it has been easy to identify a definable transport policy for the main urban areas in Britain. This has largely been for two reasons. As explained in Chapter 5, transport was, on the whole, self-financing until the 1960s. Expenditure on roads was met from the Exchequer and road users provided, and indeed still provide, one of the Treasury's main sources of income. Public transport, admittedly operating within a protected monopoly regime, broke even, although the railways were causing concern in this respect by the late 1950s. Second, there was not what we would now term a transport problem. Whilst the demand for transport as a whole had grown, notably since the 1920s, and particularly in the postwar period, there was still a belief that increasing the capacity of the transport network, and the roads in particular, would meet increasing demand. There were however a few cautionary notes. The 'Buchanan Report' entitled *'Traffic in Towns'* (Ministry of Transport, 1963) spelt out clearly the consequences for urban areas of unrestrained traffic growth. In 1964 the Smeed Committee which had been charged with examining options for what was then called road pricing, published its report which concluded that this was economically desirable and technically possible. This was another sign that unlimited growth of road traffic might not be feasible or desirable. Until this period there was, therefore, only the occasional need for government, and particularly central government, to make policy decisions on transport. The periods when they did were the early 1920s, when the railways were reorganized, the 1930s with bus service regulation, the late 1940s with nationalization, and the late 1950s and early 1960s with the crisis in railway finance. From the Transport Act, 1968, onwards however, there has been a transport policy of sorts although, more often than not, it has not been written down in a document with that title. This policy developed largely in response to growing problems of financing transport which, in turn, stemmed from the more fundamental problems which now characterize the transport policy debate.

The period before 1914

At the beginning of the century there was little need for urban transport policy as such. Apart from a few cars and carriages, all passenger transport in urban areas was public transport and it was provided commercially either by private enterprize or local authorities, who were in the process of taking over tramways. This is not to say that there were no transport problems. Congestion and pollution (albeit of a different kind to that existing today) were common in urban areas as many photographs of town and city centres at that time show. However, the laissez-faire attitude of the nineteenth century extended to transport policy. Roads were provided at public expense but the Tramways Act, 1870, allowed much of this cost to be borne by tramway operators and their passengers. Innovations on the railway network, and the introduction and subsequent electrification of tramways ensured that there was a steady increase in the supply of transport which, judging from the reductions in fares in real terms at least, was more than enough to meet rising demand as towns and cities expanded.

The inter-war years

After World War One there was substantial growth in the number of motor vehicles. This led to buses competing with and complementing trams and trains and increasing numbers of private cars. Competition was still the order of the day as far as buses were concerned, although some local authorities had exercized powers to control bus operation - usually in the interests of protecting their municipal tramways. On a national level, the competition from motorized road vehicles was eroding the competitive position of the railways and calling into question their financial stability. The railways were therefore restructured in 1923 by what amounted to compulsory mergers into four large companies which still remained in the private sector. This had the impact of reducing competition between railway companies but did little to prepare the industry for dealing with road competition.

The growth of bus usage and road haulage in the 1920s eventually led to what was regarded at the time as excessive competition and, following the Royal Commission on Transport of 1930, legislation was passed to control entry into both industries. This restricted supply and helped to reduce competition with the railways which were able to object to the granting of the licences which were now required in order to operate both bus and road haulage services. Road users were supposed to pay for the roads through an annual licence fee which was required for any vehicle to gain access to the road network. This quickly broke down when the Treasury recognized this licence as a growing source of relatively painless general revenue and latterly taxation on fuel.

Overall, the emphasis in transport policy between 1919 and 1939 was on regulating the level of competition for public transport for both passengers and freight. Subject to the protection which this regulation gave, transport operators were still able to survive commercially and therefore the level of service in urban transport was still decided by the operators - subject to the approval of the regulatory authorities. In many urban areas, buses and trams were run as trading services by the local authorities. The rest of the bus and tram industry was in private ownership although from the early 1930s onwards the railway companies diversified into ownership of buses. However, they operated their bus companies completely separately from the railways - with only token co-ordination between the two modes - which were often in competition with each other.

1945-1968

After World War Two the emphasis in the policy debate was about ownership and structure of the public transport industry. The Labour Government of 1945-50 nationalized the railways and, through their ownership of bus companies, a large part of the bus industry. The rest of the bus industry was also destined for nationalization but this was never implemented. The principle behind the policy was that a national, all-embracing transport body would offer the best and most integrated service to passengers, although, with hindsight, the problems of managing such an enterprize were severely underestimated. Further restructuring of the nationalized transport industry took place in 1953 and again in 1962 by which time the monolithic British Transport Commission (BTC) which had controlled the whole of the nationalized industry had been abolished. Railways, buses, road haulage and London Transport were now run by separate organizations.

As the finances of the railways deteriorated further, transport policy started to encompass financial issues. However, as noted in Chapter 5, no systematic attempt was made to analyze the causes of the problem until the *'Reshaping of*

British Railways' report in 1963. This only made passing reference to the underlying causes of the railways financial losses in urban areas - the uneconomic patterns of demand which caused rolling stock to be used only for very short periods of the day. However, it did not consider whether, in overall policy terms, it was more beneficial to continue to operate the services regardless of their financial performance or to accept the higher congestion which would result from their closure.

The Transport Act, 1968

As noted in Chapter 3, the election of a Labour Government in 1964 and its re-election with a substantial majority in 1966 did, however, lead to a radical review of transport policy as a whole and of urban transport in particular. A series of White Papers led to the publication of a Transport Bill in 1967 which became law as the Transport Act, 1968. This created the institutional and financial frameworks within which transport policies were to develop in the 1970s and early 1980s.

The Transport Act, 1968, is too complex to analyze in detail in this chapter but its main features were that for public transport it recognized a number of key points. First, that the conurbations were different and, as observed in Chapter 3, the then fragmented local government structure was not appropriate for decision making - hence the creation of the conurbation wide Passenger Transport Authorities (PTAs) with powers to set public transport policies for the whole of the four largest conurbations (those based on Manchester, Birmingham, Liverpool and Newcastle). The PTAs were comprized of councillors appointed by the constituent local authorities in their areas and their role was to set policies. They were given their own powers to raise revenues by levying a precept to be added to the rates of the constituent local authorities. In many respects they were the forerunners of the Metropolitan County Councils (MCCs) created in 1974. Second, that the structure of the bus and rail industry serving the conurbations was too fragmented and that there would be benefits from co-ordinating operations. Hence the creation of the Passenger Transport Executives (PTEs), whose role was to carry out the policies of their Authorities. They were given operating powers for all modes of public transport and the powers to own and operate ancillary facilities, which they used extensively. They were able to control all bus and train services in their areas - either directly or by agreement with the operators. Third, that public transport conferred wider benefits in reducing traffic congestion and therefore it should not necessarily have to be operated commercially. Powers to subsidize bus and rail services were given to the PTAs. As the local rail networks were all operating at a substantial loss and extensive closures were neither desirable in political or economic terms, these powers (under section 20 of the Act) were used extensively. Powers to subsidize bus operations were used much more sparingly in the conurbations but it was still possible to fund bus services out of passenger revenues despite there being a sound case for subsidizing bus services on wider cost-benefit criteria.

Developments since 1968

The 1968 Act, important as it was, was not the only manifestation of transport policy in the late 1960s. This was also a period when there were extensive studies of future demand for transport in the conurbations. Area wide land-use and transportation models were being developed which used the then newly developing tools of transport modelling and planning to predict future levels of car ownership and demand for all modes of transport. In policy terms, what emerged was a shopping list of investments in both highways and public transport systems which would be necessary to provide for this demand. It was therefore anticipated that transport policy would be investment led.

Again as explained in Chapter 3, the structure of local government as a whole was reorganized in 1974 and caught up with that for public transport. MCCs were created in the four conurbations which already had PTAs and also in South and West Yorkshire (based on Sheffield and Leeds respectively). The MCCs covered most of the conurbations and became the PTA as well as being responsible for the main highway network and for land-use planning and economic development. Parking could also be controlled by the MCCs although in many of them this was left to the local District Councils. This allowed, for the first time, decisions on public transport, highways, parking and land-use planning to be taken by one body - the MCCs. This reinforced the approach to transport policy which had emerged from the solutions suggested by the transport models of the late 1960s.

At the same time, transport policy started to be less financially driven. In 1974, central government grants for individual items of transport expenditure were replaced with a single grant which, in principle, could be used for any transport expenditure. At about the same time bus services started to be subsidized more heavily in the metropolitan areas and there was a growing recognition that public transport should be developed as an alternative to the private car to reduce traffic congestion and environmental impacts. As discussed in Chapter 5 there was more substantial investment in public transport, particularly rail systems. The Tyne and Wear Metro, for example, which converted a local British Rail network into a modern high frequency urban railway penetrating the centre of Newcastle by tunnel and linking both banks of the River Tyne, was opened between 1980 and 1984. In Merseyside the local rail network was expanded in the centre of Liverpool. There were also significant levels of investment in new buses and in bus stations and interchanges - much of it grant aided by central government. A different approach was adopted in South Yorkshire where public transport fares were frozen at 1976 levels. This at a time of steady, and sometimes high, levels of inflation reduced fares in real terms and started to increase patronage on the buses in particular, albeit at a cost to the rate payer.

During this period, policies to promote public transport and to restrain the use of the car were starting to be developed by the MCCs and to pay dividends - public transport patronage grew in the early 1980s in most of the conurbations. What was interesting was that there was no single policy formula - different areas had their own balance of investment and subsidy, between buses and railways, and between fares and service levels. However, there was a price to pay in terms of increasing subsidy and this gave rise to the Transport Act, 1983, which gave some indirect control of subsidy levels to central government. This was an attempt by central government to clarify what appeared to be a flaw in the legislation giving the MCCs and the Greater London Council (GLC), who were the first to be challenged in the Courts, the powers to subsidize public transport from local taxation. The Act established a mechanism whereby the Secretary of State could determine the level of subsidy which he felt was justified and protect it from legal challenge. In making his decisions, the Secretary of State could choose the criteria and in practice used social cost-benefit analysis. By this time, however, government thinking was moving away from the concepts of integrated transport systems and multi-modal transport policies. Instead, policies based on competition, deregulation and privatization were starting to come to the fore.

At the same time, rising car ownership and increasing car use, fuelled partly by land-use policies which encouraged dispersal of activity, was throwing an ever increasing volume of traffic onto the highway networks. The enginerring solution adopted by succesive governments, and explained in Chapter 6, increased the capacity of the highways by building more of them and managing traffic more efficiently but traffic congestion had continued to increase. Public transport, as a more efficient user of both urban land and road space was starting to play a larger role under the policies followed in the late 1970s and early 1980s. However, achieving this needed the MCCs and the PTAs to intervene in the market in the ways outlined above and this was clearly contrary to the policy thinking of central government.

The 1985 Acts

1985 was a landmark in the shift in policy referred to above. Two major pieces of legislation were passed that year. The first was the Local Government Act, 1985, which led to the abolition of the MCCs and the GLC, and has been discussed in Chapter 3. With most of the powers exercised by the MCCs passed to the Metropolitan District Councils, including their powers over roads, parking and land-use planning, there was no longer any authority with strategic county-wide powers and this affected transport policy. The second was the Transport Act, 1985, which introduced free competition in the provision of bus services. This deregulated the operation of bus services which since 1930 had been strictly regulated with no effective competition between operators. All this was swept away and bus operators outside of London became free to run services on a commercial basis on any route they chose. As a result some 85% of bus mileage is now provided on a commercial basis. The structure of the bus industry was changed by privatizing the National Bus Company's (NBC) operations, taking local authority bus operation out of local authority control (but not, initially at least, ownership) and by ensuring that new entrants to the market could compete for any subsidy payments which were made. As a result almost all bus services are now run by the private sector.

The implications for policy are that the scope of influence of the PTAs and their PTEs has been considerably reduced. They are now providers of the services which the market will not provide (for example impartial passenger information) and procure, under competitive contracts, bus services which are necessary to meet social needs which commercial operators find uneconomic to supply. The mechanisms by which this has been achieved are too complex to analyze in detail here but their main features are:

*   operation of public transport has been entirely separated from the bodies giving the subsidy;

*   subsidy is now for specific journeys, services etc., not for networks. Thus subsidy is at a micro-level rather than a macro level;

*   there is no subsidization of fares as a whole on the bus network;

*   decisions on bus service levels are taken by operators, not by the PTEs.

As a result, there is now very little co-ordination of bus services and there is little co-ordination between modes. On the other hand, the costs of supporting public transport have fallen as bus operators have become more efficient.

The Railways Act, 1993

Some of these principles have been carried over into the Railways Act, 1993, which will, over the next few years, start to change the nature of the local railways in the conurbations. Under this extremely complex legislation, local rail services in the conurbations will be provided by private companies operating franchizes. The franchizes will be awarded after a competitive bidding process by the PTEs and the Franchizing Director who will be jointly responsible for funding the costs of the payments to the franchize operator. Whilst it will still, in principle, be possible for the PTAs to specify services under these arrangements, there will be greater scope for the operator to influence services and fares.

Metropolitan transport problems in the 1990's

As has been seen, urban transport policy has evolved over the course of the century beginning with a period in which markets were dominant, through increasing regulation and public control and funding of transport, to a new era of competition. The most important feature of the changes which have taken place since 1985 is that the scope of influence of transport authorities in the conurbations has declined and the range of policy tools at their disposal reduced. At the same time, the problems have not been getting any less.

The most pressing transport problems in the mid 1990s, and which will still be there in the next century, concern the external impacts of transport. Using any form of transport where access to the system is not strictly controlled (as it is on railways) can lead to congestion where use of the system by one person makes all journeys slower. This was well understood in the 1960s but it has taken almost 30 years for it to get to the top of the policy agenda. Whereas in the 1960s and 1970s it was thought that increased demand for transport could be met by increasing the capacity of the network thereby relieving congestion (see Chapter 6), by the 1990s it was clear that this policy had not only failed but was environmentally unsustaintable.

The environmental concerns arise from the fact that transport imposes costs on society as a whole in the form of environmental impacts. All forms of transport create noise, pollution, severance, visual intrusion etc., although per passenger mile some are less damaging than others. Increasingly, society, has shown an unwillingness to accept the environmental consequences of transport. This has been reinforced by the growth in the demand for urban transport and the sheer scale of transport needed to serve a modern economy in the 1990s. There is also recognition among policy makers that transport, land-use and economic activity are all inter-related. The transport consequences of decisions on land-use are starting to be taken into account. The role of transport in supporting policies for land-use, and in particular urban regeneration, is also now on the policy agenda.

At the same time, many of the problems of financing transport have not gone away. By passing responsibility for public transport operation into the private sector, it has proven possible to operate a substantial proportion of the bus network (the rail network has yet to be tested in this way) on a commercial basis. This has also been achieved on parts of the London Underground and on Manchester's Metrolink light rail system. This means that the demands for subsidy have been reduced. However, the proportion of the network which can be operated commercially may not be stable in the long term and there could be a growing demand for subsidy in the future. The emphasis has, for the time being at least, switched from subsidizing operating costs to raising capital for new investment in all modes. No public transport system or road can operate commercially in the sense that revenues can cover both the costs of servicing the capital debt and its operating costs. The potential demand for capital expenditure to both maintain and expand the transport networks far exceeds the ability of the public sector alone to provide it.

The current policy response

One potential policy response could be largely based on following the market led approach which now characterizes much of transport policy. This philosophy would lead to introducing road pricing in some form or other in the conurbations which would make road users pay the true cost to the community of their journeys, including the costs they impose on other road users and on the environment. In consequence, the prices of public and private transport would reflect their relative social costs and the market would deliver the best allocation of traffic between the modes. It would also seek to capture, either through the market or by taxation, the benefits which transport confers on other interests, for example property owners and occupiers. There are many examples of this in other parts of the world - the payroll tax for transport 'Versement Transports' in urban areas in France and supplementary property taxes - Benefit Assessment Districts - in parts of the United States, for example. So far the government has not gone very far with these policies. There have been studies of the impacts of road pricing in central London and a pilot scheme for direct charging for the use of motorways is still being considered. In practice, this policy would be a radical step which needs to be tempered with reality.

The first reality is political - such a policy would mean significantly higher costs of car use in urban areas. Whilst the impact on travel demand can be predicted from transport models, the impact on votes cannot. The second reality is technological. Whilst elements of the technology to introduce road pricing (which, as has been seen, is not a new idea) exist, bringing them together as a system will be much more difficult. Recent history is littered with new, large scale, computer systems which have failed to deliver the promised output (e.g. Driver and Vehicle Licensing Centre at Swansea, Taurus on the Stock Exchange, the London Ambulance Service). In practice, such implementation is for the twenty first century. The third reality is economic.

Road pricing would not be cheap and more work needs to be done to ensure that it could deliver benefits which exceed its costs.

The market based approach to policy will not therefore be feasible for a number of years. At the same time public concern is increasingly focused upon the environmental impact of transport, transport safety (with current concerns relating to the railways), congestion and its impact upon the economy and the urban environment. In addition urban deprivation and the need for regeneration of urban areas is continuing to focus on the shortcomings and the failure of existing transport policy to address such problems.

In the early part of 1995 the then Secretary of State for Transport initiated, in a series of policy speeches, a 'Great Transport Debate' in which many of the issues discussed in this chapter were raised. In particular, he emphasized that choices had to be made and that the present situation was the result of choices which had been made in the past - for example to use the private car in preference to public transport. In 1996 the government (with a new Secretary of State for Transport in office) published a Green Paper as its response to the debate and to the report of the Royal Commission on Environmental Pollution's report on the environmental consequences of transport. The Green Paper identified the growth in road traffic as the issue at the heart of the matter, drawing attention to both its benefits and its costs. Although it covers a wider spectrum than urban transport, a number of its proposals are nevertheless particularly relevant to transport policy in the metropolitan areas.

The first of these is a series of proposals to strengthen the role of local authorities which is part of a general policy of taking decisions on local transport at local level. It is suggested that they should have wider powers to deal with vehicle-based pollution, new powers for traffic management, and play a key role in measures to increase the use of the bus. The second is continuing to develop the role of the private sector in transport operation and in funding investment in transport. Within capital expenditure programmes, there is to be a switch in emphasis from roads to public transport. Third are proposals which are a step in the direction of a more rational pricing policy, including making sure that transport users pay the full costs of their journeys - if necessary by reflecting these costs in decisions on taxation. Overall, there is a preference for pricing measures over regulation to achieve environmental improvements. Finally, there are a number of measures which are aimed at the national level to reduce vehicle emissions and improve both vehicle and personal safety.

It is too early to make any judgement on the extent to which these proposals will be taken forward into government policy. As far as the metropolitan areas are concerned, there is at least recognition that there is a problem. This can only be tackled effectively at a conurbation level on a multi-modal basis. It is accepted that car use will not fall voluntarily particularly when the quality of the existing public transport services, with some notable exceptions, falls short of that needed to attract motorists from their cars. Nevertheless there is likely to be continuing emphasis on individual choice in transport decisions fostered, where possible by competition. There is now an acceptance of the need to take action to reduce the adverse environmental impacts of transport and recognition of the inter-action between transport and land-use planning decisions. It is of paramount importance that the latter help rather than frustrate transport policy.

Implications for policy

As the Green Paper confirms, finding a way forward in these circumstances will not be easy as the existing framework and institutions are not necessarily those best equipped to deal with the problems. At the institutional level, central government has accepted proposals from local government that transport expenditure decisions should be made on a multi-modal conurbation wide basis. This is now incorporated into the package approach to transport finance in which the emphasis is on policy packages and expenditures to help achieve them rather than on individual projects. This has also had the benefit of determining policies and priorities at conurbation level rather than on a individual area basis. Whilst the government has welcomed many of the individual packages, it has allocated even less money to local transport to carry them out. At the same time, the ability of local authorities to raise money through either local taxation or borrowing is subject to stringent control by central government. Within these developments there is increasing emphasis by many local authorities on traffic restraint, improving the quality of public transport as an alternative to the car, developing non-motorized transport (i.e. walking and cycling) particularly for short distance trips. Superficially, traffic restraint seems an attractive policy. If car use can be reduced then the environmental impacts of transport and the number of traffic accidents should also go down. Increased use of public transport should also result. However, there are a number of potential problems which reduce the effectiveness of such a policy when applied on an adhoc basis by individual local authorities. One of the major concerns is the lack of a policy on traffic restraint at national level. This leaves local authorities who vigorously restrain traffic at risk of losing trade to those who do not. As there is always a proportion of economic activity which can switch easily from one urban area to another, this presents a serious risk to many urban areas contemplating widespread traffic restraint. Local authorities lack comprehensive and effective powers. For example, they only control a small proportion of parking spaces in town and city centres. Any restraint policies based on control of parking - either through price or by limitations on supply - are only partially effective in these circumstances. Again, legislation would be necessary to give local authorities control over all parking if this element of restraint is to be effective. Finally, excessive reliance on the planning system will only have impacts in the long term. In recent years the government has issued planning guidance to local authorities on a variety of matters including transport. Current guidance is that developments should only be allowed where they can be effectively served by public transport and that town and city centres - which are easiest to serve by public transport - should be the focus of development. This has been widely interpreted as the end of out of town shopping centres. However, planning permissions already granted in the past and not yet taken up will mean that further out of town developments of all kinds will continue to take place for a number of years.

Improving public transport again seems to be an attractive solution whether it is achieved by measures directed at particular modes like the bus or by a rebalancing of capital expenditure priorities. If people choose to use public transport in preference to the car then the environmental, accident and congestion problems will again be reduced because public transport is more efficient in these respects than the car. However, in the absence of either significant traffic restraint or more realistic pricing of private transport then public transport - with some exceptions - is unlikely to offer a journey time and cost which is competitive with the car. Buses, which are and will remain the dominant form of public transport, share roads with the car and are the victims and partial source of traffic congestion. As the costs of creating new rights of way are considerable in most urban areas - unless they are undergoing redevelopment - without a decision to reallocate road space in favour of the bus or to create new rights of way, the bus will rarely be as fast as the car.

Evidence is emerging, however, that this policy could work. The Manchester Metrolink light rail system offers journey times which are competitive with the car because it uses a separate right of way and has excellent penetration of the town and city centres it serves. It has attracted up to 2.5 million car journeys a year and now has a market share of well over 50% for trips between the catchment areas of its stations. Thus creating a bus service with these characteristics should have a similar impact. In practice, the quality of bus services varies considerably depending on the commercial judgements of the bus operators. Creating a high quality service will therefore need a partnership between local authorities and bus operators which not all operators may wish to join. Such a high quality service could therefore be diluted by operators running low quality services with old vehicles. A number of local authorities are trying to develop solutions to this problem although the Green Paper does not specifically address this issue in depth. Finally, there is the policy of developing non-motorized transport. A significant proportion of journeys in urban areas are less than three miles long and it is feasible to undertake them on foot or by cycle. Over the years, as the emphasis has been on improving the capacity of the road network to accommodate growing volumes of traffic, the requirements of pedestrians and cyclists have taken second priority. As a result, making journeys on foot can involve using underpasses and bridges - which may be undesirable on personal safety grounds - or long waits to cross busy roads. Similarly, until recently, little provision was made for cyclists on urban roads. Developing this policy will therefore take some time and will also involve a conscious policy decision to reallocate road space to pedestrians and cyclists. Having said this there will also be a need to allocate space between pedestrians and cyclists.

What, at the time of writing, appears to be lacking is not analysis of the problems but an acceptance that the solutions will require decisions which will not be easy to make. Greater public awareness of the problems and of the consequences of doing nothing seem to be a prerequisite for the policy decisions which are necessary. None of the solutions set out in the Green Paper - which is the only definitive statement of transport policy from the government - are without problems when it comes to implementation and there is a lot of work to do.

Transport policy in the twenty first century

As we approach the next century there is greater awareness of the importance of transport and transport policy than at any time since the early 1970s. This is most acute in the conurbations where the problems are greatest. Whilst transport policy before 1960 focused largely on the structure and ownership of the transport industry, and in the 1960s and 1970s on investment, in the 1980s and 1990s it has focused again on ownership and also on markets. At the end of the century, the focus is now shifting to the problems of meeting an ever increasing demand for transport.

There appears to be very limited prospects for major developments in transport policy in the last few years of the present century. In part, this is because the General Election which must take place by May 1997 will probably mean that controversial legislation will not be on the agenda for the remainder of the current parliament. Whilst the Green Paper recognizes the problems, it also fails to address the constraints which exist on solutions. A new government, of whichever political party, will have higher priorities than transport on its policy agenda.

In any case, experience during the twentieth century has been for an evolutionary, not revolutionary process characterizing transport policy and this seems likely to dominate at least the early years of the next century. The transport problem will not go away in the conurbations. It is essential that as we move towards the twenty first century, transport policy focuses on fundamental issues such as the objectives which transport should serve for the economy and the environment and the relative roles of each mode of transport in delivering them. Only when decisions have been made on these fundamental issues, and their consequences for transport policy understood and accepted, is there any possibility of a more rational transport policy emerging.

References

British Railways Board, (1963), *Reshaping of British Railways,* HMSO, London.
Transport, Ministry of (1963), *Traffic in Towns,* HMSO, London.

8 Conclusions

In their slim volume *'The Rise and Rise of Road Transport 1700-1990'* Barker and Gerhold set out to correct an imbalance they had detected in the transport literature against road transport in its various forms. As they put it:

> Britain's road transport has been relatively neglected ... Canals and railways, even when no longer in use, have left more visible evidence than roads and road services; and the peace of the canal towpath and the excitement of steam have attracted enthusiasts and given rise to societies and periodicals which have fostered research and a flood of publications. Road transport, by contrast, attracts few devotees and is associated all too often nowadays with frustrating traffic jams, dangerous pollution and deadly accidents (Barker and Gerhold, 1993, p.11).

Even when roads were considered by transport historians, they argued, attention was concentrated on 'turnpikes, road building and the improvement of road surfaces rather than what really mattered; the growing volume of traffic of various sorts which travelled along the roads'. As the chapters contained in the present volume show the centrality of road transport is now beyond dispute. Short of an oil crisis of cataclysmic proportions there seems little possibility that rail and coastal shipping can regain the freight transport they have lost to the roads. Similarly, it is highly unlikely that the private car, whatever the expense of motoring, will be substantially abandoned.

Although Britain has for several decades possessed an automobile culture the car has never quite achieved the level of primacy it enjoys in, for example, the United States. Nevertheless, although nobody relishes traffic jams and grid-lock; pollution and congestion; or friction and frustration; at the level of the individual vehicle the car retains considerable mystique. A car is, perhaps, the personal possession par excellence and, as such, often represents much more to its owner than merely a means of transport. Although freight movement has an obvious impact, it is the car and car ownership that lies at the heart of the urban transport problems of the industrialized world. In the urban setting cars not only replicate the services provided by public transport whether road or rail, they also provide their owners with flexibility and availability of transportation unachievable by any feasible public transport system. Thus, just as the bus, by dint of its greater flexibility, triumphed over the tram in the urban setting, so the car has more or less triumphed over the bus. Increasingly, the bus has become a residual form of transport for those who do not have access to a car.

Such is the attraction of the car that even in the central core of London where transport provision is generous, and the disincentives for drivers substantial, cars nevertheless continue to compete for available road space with buses, taxis and freight vehicles. Similar situations occur in cities throughout Britain and it is not surprizing, therefore, that many of the current debates on urban transport are couched in terms of 'wooing people away from their cars'. Indeed, much of the favourable publicity surrounding the Manchester Metrolink is concerned with the success of the system in persuading people to 'leave their car at home' and use the supertram instead. Interestingly, very few commentators anticipate the total abandonment of the car in favour of a return to total dependency on public transport. Instead, the hope is that a significant number of car owners will accept a public transport substitute for all, or at least part, of some of their journeys. Of course, if this is to occur, then the quality of journey experienced on public transport must equate to that provided by the private car. Alternatively, and perhaps more controversially, a local transport regime must be established which prohibits private cars from certain traffic corridors and city zones.

The chapters contained in this volume have demonstrated that many urban transport problems are not amenable to easy solutions. Providing improved urban transport infrastructure is expensive and can be environmentally hostile. Obviously the interests of local residents should be balanced against those of commuters, tourists or those merely 'passing through'. In the context of this debate the prominence historically given to the ownership of transport (whether public or private) in political controversy on urban transport is becoming less relevant. Instead a partnership approach offers perhaps the best way forward, involving affected communities, user groups and providers in the decision-making process. Similarly, finding the financial wherewithal to bring about improvement cannot rest with either the state (whether supra, national or local) or private capital alone. Again, a partnership approach offers the most likely way forward, balancing the needs for investment with the possibilities of gaining a viable return on capital. Of course, such an approach tends to be piecemeal, pragmatic and largely dependent on local knowledge and experience. However, having said this, it is important to acknowledge that 'localism' has its limitations and may become constrained by narrow sectional considerations. Thus a policy framework must be developed by central government which can at least prompt local decision-making in a 'progressive' direction. Again a partnership approach is likely to provide the best way forward. As the government's 1996 Green Paper put it, 'central government must not only provide the right policy framework, but must also be prepared to offer support, to share expertise, to ensure the necessary powers are available, and promote a national vision of a quality environment' (Department of Transport, 1996, p.106).

References

Barker, T. and Gerhold, D. (1993), *The Rise and Rise of Road Transport 1700-1990*, MacMillan, London.

Transport, Department of (1996), *Transport the Way Forward*, Cmnd.3234, HMSO, London.

Bibliography

Artibise, A. and Stetler, G. (eds.) (1979), *The Usable Urban Past: Politics and Planning in The Modern Canadian City*, MacMillan, Toronto.

Alexander, A. (1982), *Local Government in Britain since Reorganization*, Allen & Unwin, London.

Bagwell, P. (1986), *The Transport Revolution 1770-1985*, Routledge, London.

Barker, T. (1990), *Moving Millions: A Pictorial History of London Transport*, London Transport Museum, London.

Barker, T. and Robbins, M. (1963), *A History of London Transport: Volume 1 - The Nineteenth Century*, Allen & Unwin, London.

Barker, T. and Robbins, M. (1976), *A History of London Transport: Volume 2 - The Twentieth Century to 1970*, Allen & Unwin, London.

Barker, T. and Gerhold, D. (1993), *The Rise and Rise of Road Transport 1700-1990*, Macmillan, London.

Bell, P. and Cloake, D. (1990), *Deregulation and Transport Forces in a Modern World*, Fulton, London.

Birks, J. (1990), *National Bus Company 1968-89: a commemorative volume*, Transport Publishing Company, Glossop.

Bosley, P. (1990), *Light Railways in England and Wales*, Manchester University Press, Manchester.

British Railways Board, (1963), *Reshaping of British Railways*, HMSO, London.

British Railways Board, (1992), *Annual Report and Accounts 1991-92,* HMSO, London.

British Transport Commission, (1961), *Annual Report and Accounts 1960,* HMSO, London.

Byrne, T. (1992), *Local Government In Britain,* Penguin, Harmondsworth.

Castle, B. (1990), *The Castle Diaries 1964-76,* Papermac, London.

Chandler, J. (1991), *Local Government Today,* Manchester University Press, Manchester.

Cherry, G. (ed.) (1980), *Shaping an Urban World,* Mansell, London.

Confederation of Passenger Transport, UK, (1995), *Park and ride bus services: opportunities and prospects,* (unpublished), London.

Confederation of Passenger Transport, UK, (1996), *Fuel duty rebate: submission to HM Treasury,* London.

Donoughue, B. and Jones, G. (1973), *Herbert Morrison Portrait of a Politician,* Weidenfeld & Nicolson, London.

Dreiser, T. (1912), *The Financier,* Harper, New York.

Dreiser, T. (1914), *The Titan,* John Lane, New York.

Dreiser, T. (1947), *The Stoic,* Doubleday, New York.

Dyos, H. and Aldcroft, D. (1971), *British Transport,* Leicester University Press, Leicester.

Environment, Department of (1976), *Transport Policy: A Consultative Document,* HMSO, London.

Environment, Department of (1983), *Streamlining the Cities,* HMSO, London.

Gerecke, K. (ed.) (1991), *The Canadian City,* Black Rose Books, Montreal.

Glover, J. (1996), *London's Underground,* 8th edition, Ian Allan, London.

GO Transit, (1992), *Twenty Five Years on the Go,* GO Transit, Ontario.

Green, O. (1990), *Underground Art: London Transport Posters 1908 to the Present,* Studio Vista, London.

Hampton, W. (1991), *Local Government and Urban Politics,* Longman, Harlow.

Hibbs, J. (1968), *The History of British Bus Services,* David & Charles, Newton Abbot.

Higginson, M. (ed.) (1987), *Tramway London: background to the abandonment of London's trams 1931-52,* Light Rail Transit Association, Birkbeck College & London Transport Museum, London.

Holt, G. (1978), *A Regional History of the Railways of Great Britain, Volume 10: The North West,* David & Charles, Newton Abbot.

Jackson, A. (1973), *Semi-detached London: Suburban Development, Life and Transport 1900-39,* Allen & Unwin, London.

Jackson, A. (1978), *London's Local Railways,* David & Charles, Newton Abbot.

Jackson, A. (1986), *London's Metropolitan Railway,* David & Charles,

Newton Abbot.

Jackson, A. (1991), *Semi-Detached London: Suburban Development, Life and Transport 1900-39*, 2nd edition Wild Swan Publications, Didcot.

Jacobs, J. (1965), *The Death and Life of Great American Cities*, Penguin, Harmondsworth.

Kain, R. (ed.) (1981), *Planning for Conservation*, Mansell, London.

Kingdom, J. (1991), *Local Government and Politics in Britain*, Philip Allan, Hemel Hempstead.

Klapper, C. (1978), *Golden Age of Buses*, Routledge & Kegan Paul, London.

Leach, S. (ed.) (1994), *The Local Government Review: Key Issues and Choices*, Institute of Local Government Studies, University of Birmingham, Birmingham.

Leach, S. Davis, H. Game, C. and Skelcher, C. (1990), *After Abolition: The Operation Of The Post-1986 Metropolitan Government System In England*, Institute of Local Government Studies, University of Birmingham, Birmingham.

Lee, C. (1973), *The Piccadilly Line: a brief history*, London Transport, London.

Lemon, J. (1985), *The History of Canadian Cities: Toronto Since 1918*, Lorimer & National Museums of Canada, Toronto.

London First Transport Initiative for the London Pride Partnership, (1996), *London's action programme for transport: 1996-2010*, London.

Lubove, R. (1967), *The Urban Community: Housing and Planning in the Progressive Era*, Englewood Cliffs, New Jersey.

McConville, J. (ed.) (1997), *Transport Regulation Matters*, Cassell, London.

McConville, J. and Sheldrake, J. (eds.) (1995), *Transport in Transition*, Avebury, Aldershot.

Merseyside Passenger Transport Executive and British Rail, (1978), *The story of Merseyrail*, Liverpool.

Metropolitan Toronto, (1992), *The Liveable Metropolis: Metropolitan Toronto Official Plan* (Draft), Toronto.

Metropolitan Toronto, (1994), *The Official Plan of the Municipality of Toronto*, Toronto.

Morris, O. (1953), *Fares Please: the Story of London's Road Transport*, Ian Allan, London.

Morrison, H. (1933), *Socialization and Transport: The Organization of Socialized Industries with particular reference to the London Passenger Transport Board*, Constable, London.

Munby, D. (1978), *Inland Transport Statistics Great Britain 1900-70*, Clarendon Press, Oxford.

Nowlan, D. and N. (1970), *The Bad Trip: the Untold Story of the Spadina Expressway*, New Press/House of Anasi, Toronto.

Official Report, Fifth Series, (1966-67), *Parliamentary Debates*, Volume 746, London.

Pearce, C. (1980), *The Machinery of Change in Local Government 1888-1974*, Allen & Unwin, London.

Pearce, M. and Stewart, G. (1992), *British Political History 1867-1990: Democracy and Decline*, Routledge, London.

PTE, Tyneside (1973), *Public Transport On Tyneside: A Plan For The People*, Tyneside Passenger Transport Authority, Newcastle upon Tyne.

Redcliffe-Maud, Lord (Chairman) (1969), *Report of the Royal Commission on Local Government in England 1966-69*, Cmnd.4040, HMSO, London.

Redcliffe-Maud, Lord and Wood, B. (1974), *English Local Government Reformed*, Oxford University Press, London.

Richards, P. (1980), *The Reformed Local Government System*, Allen & Unwin, London.

Royal Commission on Transport, (1931), *Final Report*, HMSO, London.

Savage, C. (1952), *An Economic History of British Transport*, Heinemann, London.

Savage, C. (1966), *An Economic History of Transport*, Hutchinson, London.

Sewell, J. (1993), *The Shape of the City*, University of Toronto Press, Toronto.

Shaw, S. (1993), *Transport: Strategy and Policy*, Blackwell, Oxford.

Sheldrake, J. (1996), *Management Theory: Taylorism to Japanization*, International Thomson, London.

Sherrington, C. (1969), *A Hundred Years of Inland Transport*, Frank Cass, London.

Smeeton, C. (1994), *The London United Tramways: Volume 1 - Origins to 1912*, LRTA/TLRS, London.

Stanyer, J. (1976), *Understanding Local Government*, Martin Robertson, Oxford.

Thomas, D. (1970), *London's Green Belt*, Faber, London.

Thomas, H. and Krishnarayan, V. (eds.) (1994), *Race Equality and Planning*, Avebury, Aldershot.

Thomas, J. revised by Paterson, A. (1984), *A Regional History of the Railways of Great Britain, Volume 6: Scotland - the Lowlands and the Borders*, David & Charles, Newton Abbot.

Thomson, J. (1974), *Modern Transport*, Penguin, Harmondsworth.

Tomlinson, W. (1987), *The North Eastern Railway: its rise and development*, David & Charles, Newton Abbot.

Toronto Transit Commission, (1989), *Back to Basics: A TTC Strategy for the 1990s*, Toronto.

Toronto Transit Commission, (1989), *Moving Forward: Making Transit Safer for Women*, Toronto.

Toronto Transit Commission, (1989), *Transit in Toronto: the story of public transportation in Metropolitan Toronto*, Toronto.

Toronto Transit Commission, (1992), *Ethno-Racial Access Plan*, Toronto.

Traffic Director for London, (1993), *Network Plan Consultation Document*, London.

Traffic Director for London, (1996), *Annual Report 1995-96*, London.

Transport, Department of (1983), *Railway Finances*, HMSO, London.

Transport, Department of (1995), *Transport Statistics Great Britain, 1995 Edition*, HMSO, London.

Transport, Department of (1996), *Transport the Way Forward*, Cmnd.3234, HMSO, London.

Transport, Ministry of (1949), *British Transport Commission London Plan Working Party, Report to the Minister of Transport*, HMSO, London.

Transport, Ministry of (1963), *Traffic in Towns*, HMSO, London.

Transport, Ministry of (1966), *Transport Policy*, Cmnd.3057, HMSO, London.

Transport, Ministry of (1967), *Public Transport and Traffic*, Cmnd.3481, HMSO, London.

Travers, T. and Glaister, S. (1994), *An Infrastructure Fund for London*, Greater London Group, London School of Economics, London.

Urwick, L. and Brech, E. (1953), *The Making of Scientific Management Volume 2 - Management in British Industry*, Pitman, London.

Warren, K. (1980), *Fifty Years of the Green Line*, Ian Allan, London.

Wilson, W. (1989), *The City Beautiful Movement*, The John Hopkins University Press, Baltimore.

Wiseman, R. (1986), *Classic Tramcars*, Ian Allan, London.

Wood, B. (1976), *The Process Of Local Government Reform 1966-74*, Allen & Unwin, London.

Index

Evans, Bob 61
expressways - see roads

Fascist Italy 12
Feltham, Middlesex 15
ferries 101
Finance Act 1965, 85
Ford, G. B. 50
Formby 31
freight 1, 5, 8, 48, 52, 72-74, 94-95, 106, 118-19
Freightliner 73

Gardiner, F. 56
Gateshead 30
GEC Alsthom 82
Gibb, Sir George 16
Glasgow 5, 71-72, 76, 79, 84
Glasgow Corporation 84
Glasgow District Subway Company 79
Glossop 31, 34, 38
Gluckstein, S. 24
goods transport - see freight
Go Transit 58-60
Great Eastern Railway 72, 74
Great Northern Railway 72, 79
Great Western Railway (GWR) 81
Greater London 23
Greater London Council (GLC) 23, 35, 109
Greater Manchester 38-39, 43
green belt 81-82
Green Line 15
Greenwich 74, 83
grid-lock 1, 119
Guild Socialism 20

Halifax 30, 74
Harrogate 39
Harvey, D. 53-54
Hastings, Dr. C. 52

haulage - see freight
Heathrow Express 76
Henderson, A. 21
Herbert, Sir Edwin 35
highways - see roads
Hocken, H. 52-53
Home Counties 12
horses 12-13, 78
Horwich 34
Howland, W. 51
Huddersfield 30
Hull, Canada 50

Ilford 23
Immigration Act 1967 (Canada) 64
industrial democracy 26
International Monetary Fund 99-100
investment 4-6, 70-72, 75-84, 87-89, 108, 120
Ipswich 79

Johnston, E. 17

Kansas City 49
Keighley 74
Knattries, Albert - see Ashfield, Lord
Knutsford 43

Labour Government and Party 2-3, 8, 18-22, 25, 31-32, 37, 40, 99-100, 106-7
Lancashire 37, 39
Leach, A. 62
Leeds 30, 74, 77, 81, 108
Leigh 34-35
Leyton 23
Lichfield 35
light Railways - see railways
Light Railways Act 1896, 73-74
Liverpool 5, 13, 30, 72, 76, 79, 83-

84, 95, 107-8
Liverpool City Council 83-84
Liverpool Overhead Railway
  (LOR) 83-84
Lloyd George, D. 12, 16
local authorities and government
  3-4, 6, 8-9, 19-20, 23, 28-44,
  46-47, 50-56, 59, 61, 66-67, 74-
  79, 84, 86-87, 105-10, 113-16
local democracy - see local
  authorities and government
Local Government Act 1958, 30
Local Government Act 1972, 42,
  86
Local Government Act 1985, 40,
  109
Local Government Act 1992, 42
Local Government Boundary
  Commission 30, 42
Local Government (Boundary
  Commission) Act 1945, 30
Local Government Commission for
  England 30-31, 33-34, 42,
local government Special Review
  Areas (SRAs) 30-31, 33-35, 43
London 2-3, 5, 10-16, 18, 20, 23-
  24, 35, 71-72, 74, 76-88, 101,
  110-12, 119
London, Chatham and Dover
  Railway 74
London and Home Counties Traffic
  Advisory Committee 18, 21-22
London and North Eastern Railway
  (LNER) 72, 81
London and South Western
  Railway 79
London and Suburban Traction
  Company 14
London County Council (LCC) 2,
  10-11, 13-15, 18-19, 21, 23, 81
London County Coucil (Co-
  ordination of Passenger Traffic)

Bill 1929, 18-19
London Electric Railways (Co-
  ordination of Traffic) Bill 1929,
  18-19
London General Omnibus Company
  (LGOC) 2, 15-17, 80
London Government Act 1963, 23
London, Midland and Scottish
  Railway 83
London Passenger Transport Act
  1933, 23
London Passenger Transport Board
  (LPTB) 2-3, 5, 10, 12, 14-17, 19,
  22-26, 29, 81
London Regional Transport 82
London Traffic Act 1924, 13-16,
  18-19
London Traffic Area 1924, 19
London Transport 12, 17, 80-82
  106
London Transport (Finance) Act
  1935, 25
London Underground (LUL) 79,
  82-83, 111
London United Electric Tramways
  (LUT) 14-15, 78-80
Longdendale 31
lorries 95

MacDonald, R. 21
Mackenzie, W. 52
Macmillan, H. 12
mainline railways - see railways
Manchester 13, 30, 87, 107
Manchester Metrolink 76, 79, 83,
  112, 115, 119
Marsh, B. C. 49
Massachusetts Bay Area
  Transportation Authority 32
Menzler, F. A. A. 17
Meriden 39
Merseyrail Electrics Limited 83

Mersey Railway 79, 83
Merseyside 28, 30-35, 38-39, 43,
  83-84, 108
Merseyside Docks and Harbour
  Board (MDHB) 83-84
Metrac 62-63
Metroland 72
Metropolitan Board of Works 13
Metropolitan County Councils
  (MCCs) - see local authorities and
  government
Metropolitan District Railway 24,
  79
Metropolitan Electric Tramways
  (MET) 14-15
Metropolitan Police District 15
Metropolitan Railway 19, 24, 72,
  79
Metropolitan Toronto (and Metro)
  55-67
Metropolitan Water Board 13
Midland Metro 87
Midland Railway 74
mobility 6-7, 11, 61, 93, 96, 100
Morrison, H. 2, 18-23, 25-26
motor buses - see buses
motor cars and motoring - see
  cars
motor cycles 96-97
motorways - see roads
municipal and municipalization -
  see local authorities and
  government
municipal pride - see civic pride

National Bus Company (NBC) 84-
  85, 110
National Coalition Government 2,
  21, 24
nationalization - see public
  ownership
Nazi Germany 12

Neston 38
New Jersey 16
New Mills 31, 35
New York 16, 49
Newcastle 30, 76, 83, 87, 95, 107-8
North Eastern Railway Company
  16, 73
North Woolwich 75
Nottingham 74

Oldham 30
omnibuses - see buses
Ontario 51, 55-56, 58-61
Organisation for Economic Co-
  operation and Development
  (OECD) 94
Ottawa 50
Oxford 75, 87

Paris 16
park and ride - see parking
parking 1, 6, 55, 58, 87-88, 93,
  108, 110, 114
passenger transport areas 3, 28-29,
  32-36, 38-44
Passenger Transport Authorities
  (PTAs) 32, 34, 38, 40, 86, 107-
  11
Passenger Transport Executives
  (PTEs) 5, 33, 75-76, 79, 83-86,
  107, 110-11
pedestrians 1, 114, 116
Philadelphia 80
Pick, F. 12, 16-23, 25
Port of London Authority 13
ports 51, 83, 94
pollution 1, 88, 92, 102, 105, 111,
  113, 119
Potter, B. 60
Poynton, 39
Prescott 31
privatization 5-9, 41, 76-77, 79, 82,